Power Corrupts

Hilary Bacon and John Valentine

Power Corrupts

The Arguments Against Nuclear Power

 Pluto Press

First published 1981 by Pluto Press Limited,
Unit 10 Spencer Court, 7 Chalcot Road, London NW1 8LH

Copyright © Hilary Bacon and John Valentine

ISBN 0 86104 345 6

Cover photograph by Michael Abrahams
Cover design by Clive Challis

Typeset by Grassroots Typeset, London.
Printed in Great Britain by Blackwell Press Limited.

Contents

Do you think you can take over the universe and improve it?
I do not believe it can be done.

The universe is sacred.
You cannot improve it.
If you try to change it, you will ruin it.
If you try to hold it, you will lose it...

Therefore the sage avoids extremes, excesses, and complacency.

Lao Tsu

We should like to thank all those friends and fellow-thinkers without whose generous help and work this book could not have been achieved. Thanks are also due to Peter Bunyard, Dr Helen Caldicott, Shoja Etemad, Daniel Pisello and Richard Webb, from whose writings we have quoted extensively.

Introduction

Nuclear power stations, apart from the hazards they create, produce only electricity. Electricity alone cannot satisfy our present or future energy needs, and cannot easily be stored. During the process of generating electricity, nuclear power stations also generate highly poisonous radioactive waste, a great deal of which has already been dispersed into our environment. The radiation from this waste causes cancer, genetic mutations and species degeneration in all living creatures. No safe method has been found for disposing of this waste. It cannot be destroyed. It lasts for thousands of years. Research into this urgent problem is only one of the reasons why millions of pounds of taxpayers' money have been poured — unproductively — into this industry, which so far generates only 13 per cent of our electricity supply ('File on 4', BBC Radio 4, 12 December 1979).

Why has the nuclear mountain laboured so expensively and so dangerously to bring forth this mouse? The answer lies in the military connection. The hidden principal product of nuclear power stations is plutonium, the lethal radioactive element used in nuclear weapons and aptly named after the king of the underworld. The production of military-use plutonium was the original aim of the government when the first nuclear reactor was built at Calder Hall in 1956-57. Electricity was simply a by-product. Plutonium is the reason why this highly specialised industry has operated amid secrecy from the very beginning and also explains the unlimited financial, technological and ultimately political power granted to it by successive governments.

Yet even with the present relatively small UK nuclear power programme, the country already has at least 7.5 tonnes of plutonium in excess of its military demands. This cannot be disposed of (except possibly to countries anxious to develop their own nuclear weapons), and about 250,000 years will have to elapse before it decays naturally into harmlessness. But it can be used to produce *more* plutonium in the highly explosive and technologically unproved fast-breeder reactor (FBR) concerning whose development the present government has expressed enthusiasm.

On 23 October 1979, a Cabinet meeting was held at which

ministers agreed to pursue a greatly expanded nuclear power programme, but without informing the public too clearly, for fear of confrontation (*Time Out, Guardian*). In the *Observer* of 29 June 1980, Adam Raphael reported that a government meeting planned to consider a £4,000 million 20-year programme for the development and construction of a commercial fast-breeder reactor. The meeting hoped to find ways of co-operating with other countries (notably France or the US) in order to minimise costs. However, President Carter had already ordered a moratorium on the construction or use of FBRs because of their dangers, and France is demanding £50 million as an initiation fee into its programme. A further point, which the Department of Energy does not mention, is that an international link-up would effectively silence opposition within the UK to this programme; nor has the Secretary of State for Energy disclosed where or when the first FBR will be built in Britain. Adam Raphael notes: 'The 20-year programme has the enthusiastic support of the Prime Minister, but is running into opposition from the Treasury because of its enormous price tag.' In the face of such covert undertakings which radically affect all our lives, it is clear that the public must be informed of the cogent and well-documented arguments for opposing any type of nuclear programme.

Almost all the arguments against the civil use of nuclear power were exemplified by its first devastating military use in the atomic bombs which destroyed Hiroshima and Nagasaki. The military situation demanded secrecy; thus the nuclear technology which had been developed so quickly was understood by only a handful of physicists, and its effects on living creatures apparent to only a handful of biologists. The politicians who took the decision to drop the bomb, among them Truman and Attlee, claimed that they had not been fully informed of the radiation effects of nuclear fission when they made their decision. This combination of a new and dangerous technology being developed before it has been completely understood or publicly accepted, supra-normal financial and political power concentrated in a few 'expert' hands, and long-term hazards which only subsequently manifest themselves, is unfortunately still characteristic of the civil use of nuclear power to generate — however ineffectively — electricity.

Glossary

It might be helpful to explain some terms and abbreviations used in the text:

Magnox: the early British design of gas-cooled reactor. There are 5 in service, but many are nearing the end of their useful lives. It is worth pointing out that when the CEGB make cost comparisons of electricity generation using coal-fired, oil-fired and nuclear methods, they quote the Magnox reactors as representing the nuclear cost. The resulting comparisons are favourable to the nuclear lobby, since there are no outstanding R & D costs for these reactors, and their fuel, used at relatively low burn-up temperatures, does not have to go through the very expensive fuel enrichment process necessary for the more advanced types.

AGR: the later British-designed Advanced Gas-Cooled Reactor. There are three in service and, although they are probably inherently safer than American reactors, they have been bedevilled by technical failures. As Sir Arthur Hawkins, ex-chairman of the CEGB, said: 'The AGR programme is a disaster we must not repeat.'

PWR: the American Pressurised Water Reactor, originally developed to power nuclear submarines. A large number of these reactors are in service world-wide — the best-remembered (but possibly not the most serious) accident involving this type of reactor occurred at Three Mile Island in March 1979. It seems certain that the present government intends PWRs to play a significant part in the installation of ten new reactors in the next decade.

All three types above are 'thermal' reactors: they use uranium or uranium oxide as fuel and produce, among other substances, plutonium as a waste product.

FBR: the Fast Breeder Reactor may be regarded as the next generation of nuclear reactor. FBRs appear to provide a neat technological answer to the problems of uranium scarcity and undisposable waste plutonium from the thermal reactors, in that they can use plutonium as fuel and breed more plutonium than they consume. They are potentially vastly more dangerous than

9

the thermal reactors because of the huge amounts of plutonium they contain and the consequent possibility of an atomic explosion. FBRs operate at much higher temperatures than thermal reactors, which means that they are closer to the limits of tolerance of both men and technology at every stage of their activity. The only coolant effective at these temperatures is liquid sodium, one of the most chemically reactive substances known. Liquid sodium reacts explosively with both air and water, as the recent explosion at the British experimental FBR at Dounreay graphically demonstrated.

Reprocessing: in thermal reactor terms, this involves transporting the spent fuel from reactors all over the country to Windscale in Cumbria, extracting the plutonium and other waste materials, recycling the unused uranium, storing the plutonium for 250,000 years (or until an effective disposal method is found, whichever is the sooner) and disposing of the other radioactive substances according to their toxicity. Reprocessing of fast breeder reactor (FBR) fuel is still in the design and experimental stage, and is a much more critical process. This is because of the substantially larger quantities of plutonium involved, both unconsumed in the spent fuel and bred in the reactor.

Enrichment: this very expensive process takes place at Capenhurst, and enriches the uranium oxide fuel so that it is suitable for use in the more advanced types of thermal reactors. An enrichment plant in Ohio uses 10 per cent of that state's electricity — more than the city of Cleveland.

High Burn-up: the AGR and PWR are both high burn-up reactors. This means that their fuel rods, once loaded, remain operative in the reactor longer than Magnox fuel rods, which poison themselves more quickly with radioactive fission products that slow down the protons and therefore reduce heat production. However, there are difficulties associated with high burn-up fuel rods. They are more dangerous to reprocess than Magnox fuel rods, being far more radioactive and also thermally hotter to handle.

Curie: unit of radiation indicating 37 billion radio-active disintegrations per second.

1. Radiation

For many years now, the nuclear industry has been claiming that it is a safe industry. The nuclear safety question has three aspects: routine low-level radioactive emissions; high-level radioactive waste; and leaks and accidents within the nuclear plants themselves. The industry claims that those routine low-level emissions which disperse into the environment are safe because they have been kept within the 'permissible' limits set up by the International Commission for Radiological Protection. Yet these same limits have been changed several times over the last few years, and the whole idea of 'permissible' limits for ionising radiation has been questioned by both doctors and biologists.

Even before the production of the first atomic bomb, it was known that radiation such as that routinely produced by nuclear power stations causes cancer, genetic mutations, and a reduction in immunity to disease. Studies of the effect of ionising radiation on genes were begun in 1925, and summed up in 1957 by the geneticist Theodore Dobzhanski, in an article in *Science*: 'if anything, radiation-induced mutants are more destructive than spontaneous ones. As far as genetic effects are concerned, the only safe dose of high-energy (i.e. ionising) radiation is no radiation at all' ('Genetic loads in natural populations', *Science* 126, 1957). Now, in the 35 years since Hiroshima, during which period the nuclear industry has been expanding rapidly and dispersing large quantities of ionising radiation into our environment, scientists have been studying the effects of even low-level ionising radiation on living creatures, its concentration in the food chain, and the statistical spreads of leukaemia, lung cancer and other associated damage near radiation sources. Since these effects can take up to 30 years to reveal themselves, the industry has been well served by this time-lag; now, however, it is clear that the claims of nuclear safety must be challenged.

The question of nuclear radiation causing mutations, the slow degradation of whole species and their possible extinction, was raised by Hermann Muller and finally published in the *Journal of the American Public Health Association* in 1964. This being so, how can the American National Council on Radiation

Protection and the International Commission on Radiological Protection have prescribed a 'permissible' exposure limit, when the public has not been asked whether it gives its permission for the presence of such hazards within the environment? The answer is given in part by an eminent scientist from within the nuclear establishment, Dr Karl Morgan, former chairman of the American National Council on Radiation Protection and of the International Commission on Radiological Protection. Dr Morgan worked for many years at the US Atomic Energy Commission laboratory at Oak Ridge, Tennessee, and was the founder and director of the Health Physics division there from 1943 to 1972. On 27 October 1979 a conference on low-level ionising radiation was held at Guy's Hospital, London, sponsored by the Unit for the Study of Health Policy, Department of Community Medicine. Dr Morgan spoke at this conference:

> During the first years of the Atomic Age (1942-60) a large number of scientists (perhaps most who were knowledgeable in health physics and radiobiology) accepted the threshold theory that there is a safe level of exposure to ionising radiation, and that as long as a person does not exceed this threshold or safe level no harm will result, or the radiation damage will on average be repaired as fast as it is produced. From 1960 to the present, an overwhelming amount of data have been accumulated that show there is *no safe level* of exposure, and there is no dose of radiation so low that the risk of malignancy is zero. Therefore, the question is not: Is there a risk from low-level exposure? Or: What is a safe level of exposure? The question is: How great is this risk? Or: How great may a particular radiation risk be before it exceeds the expected benefits, such as those from medical radiography or nuclear power? It is obvious to all scientists in the field, as well as to diehards for the threshold hypothesis, that at least for some types of radiation damage and for some kinds of radiation exposure (especially from low LET radiation, that is, X, gamma and beta radiations) there is some repair of the radiation damage going on in the body. The diehards, however, do not seem willing or able to accept the evidence that for man there is never a complete repair of the radiation damage, since even at very low exposure levels there are many thousands of interactions of the radiation with cells of the human body ...
>
> There are undoubtedly many mechanisms of radiation injury such as damage to cell membranes, damage to the body repair mechanisms, indirect damage (for example, damage to cell blood supply and formation of harmful chemicals such as hydrogen

peroxide in cell cytoplasm) and impairment of efficiency to lung clearance mechanisms. Each of these mechanisms may contribute to the development of a malignancy. However, perhaps the most significant damage from low-level exposure results from direct interaction of the stream of ions with the nucleus of one of the billions of irradiated cells that may in the rare event survive and continue to divide but fail to repair the radiation damage. There are 46 chromosomes in the nucleus of each normal somatic cell of the human body, and along each chromosome are coded millions of bits of information like an immense library which enables or instructs the cell to function properly and to divide or stop dividing at the appropriate time. When radiation passes through the human body, four principal events can occur:

> The radiation pases through or near the cell without producing any damage.
>
> The radiation kills the cell or renders it incapable of cell division.
>
> The radiation damages the cell but the damage is repaired adequately.
>
> The cell nucleus (or library of information) is damaged but the cell survives and multiplies in its perturbative form over a period of years (5 to 70 years) and forms a clone of cells that eventually is diagnosed as a malignancy.

Only this last event relates to somatic damage such as cancer from low-level exposure. It is obvious that if the cell nucleus is damaged and some information is lost or if a similar series of events leads to the development of a malignancy, there can be no dose so low that the risk is zero.

Another equally eminent scientist from within the nuclear establishment is Dr John Gofman, a former US government scientist who first identified the isotope uranium-233. He is now professor emeritus of medical physics at the University of California in Berkeley, and chairman of the Committee for Nuclear Responsibility. He has described the process and the results of ionising radiation in his book *Poisoned Power*:

> Ionising radiation (from beta-rays, X-rays, gamma-rays or alpha particles) rips negatively-charged electrons from atoms, leaving positively-charged ions ... some radiation, travelling nearly at the speed of light, has enough energy to break 100,000 chemical bonds between atoms ... If the damage is catastrophic, the cell which has experienced the radiation injury dies. If less than that, the cell can go on living, though wounded, for a long time. Not only can wounded cells go on living, they can divide and reproduce new

cells. Unfortunately, these new cells might carry the injury sustained by the irradiated cell from which they originate ... Non-fatal injury to the cells of certain human tissues may be far, far more dangerous to the person than the outright, immediate death of the cell would be ... because, within a period of years, a single cell injured in this way has the potential to initiate a cancer or a leukaemia ... We do know for certain that this process does occur ... but once this process has been initiated by radiation, science knows of no way to stop it ... The period between radiation injury and obvious cancer is quite long in the human. For leukaemia, the latency period is some five years; for thyroid cancer, it is approximately thirteen years. For some other cancers, the latency period is still not accurately known, although periods of twenty years or more have been suggested.

In May 1976 the National Academy of Sciences published a report, 'Risks associated with nuclear power'. This report estimated that by the year 2000, about 2,000 people would die of cancer *as a result of radiation leakage*, without assuming any particular disasters.

The step-by-step health hazard effects of the various radioactive elements produced during the nuclear fuel cycle have been clearly and urgently detailed by Dr Helen Caldicott, who is president of the American group of doctors, Physicians for Social Responsibility. In her paper, 'What Physicians Should Know About Nuclear Power', she sums up these radioactive processes within the human body:

The fuel cycle of nuclear power plants is complex, but not too difficult to understand. It has many biological and medical implications which must be understood by physician and patient, as well as by the politicians who make most important decisions for society. In this article, I will describe the fuel cycle step-by-step and explain the medical dangers arising from each step.

Mining. When uranium — the fuel for atomic reactors — is mined from the ground it emits the radioactive gas called *radon*. Often inhaled into the lungs of miners, after four days radon converts to *lead-210* which remains radioactive for more than 100 years. Because radiation in the body is carcinogenic, it has been discovered in the United States that up to 20 per cent of uranium miners die of lung cancer over a 20-year period of mining.

Milling. After the uranium ore is mined, it is then milled and refined. Thousands of tons of waste ore (tailings) are discarded and left lying in huge heaps on the ground. The gas radon is continually emitted from the waste uranium in the tailings. The tailings generated to provide uranium for nuclear power in the United States over the next twenty-four years may produce, through the causative agent radon, 45 cases of lung cancer in the world per year for tens of thousands of years.

Enrichment and fuel fabrication. The uranium is then enriched and fabricated into fuel rods which are transported to the nuclear reactor and placed in the reactor core. A typical 1,000 megawatt reactor contains 526 bundles, each bundle consisting of 12 rods. The radioactive uranium produces heat by fission which is utilised to generate electricity. But during this process uranium is converted to many radioactive by-products which are the ashes or wastes of nuclear power. Once a year one-quarter of the rods are removed from the reactor core because their generating life has ceased. The rods are both thermally and radioactively very hot and must be stored on racks in cooling ponds containing water. They now contain a very large number of biologically dangerous radioactive materials, including strontium-90, iodine-131, cesium-137 and plutonium. Eventually it is hoped that these rods will be transported in caskets to a reprocessing plant where they will be dissolved in nitric acid.

Reprocessing. The plutonium is purified and removed from the solution in powder form as plutonium dioxide. It will then be used as either fuel for atomic bombs or fuel for 'fast breeder reactors' (reactors which breed plutonium). It is at this point in the fuel cycle that the greatest dangers arise once the plutonium is separated.

Plutonium is an extremely potent cancer-producing material, appropriately named after Pluto, the God of the dead and ruler of the underworld. It enters the body by inhalation of contaminated air and is deposited in the lungs. Because of its potent cancer-producing properties, the acceptable body dose has been set at less than 1 millionth of a gram (an invisible particle). There is some evidence that this level has been set too high. Cancer will not appear until 15 to 20 years after inhalation. By extrapolation, 1lb of plutonium, universally dispersed, would be adequate to kill every man, woman and child on earth.

Most of the plutonium manufactured in the fuel cycle will be in powdered form and by the year 2020 in the United States the industry will have produced 30,000 tons of plutonium and there will be 100,000 shipments of material annually on American highways. Because plutonium is the basic material of atomic bombs, it is more valuable than heroin on the black market, therefore vulnerable to theft by terrorists, racketeers, non-nuclear nations and deranged individuals.

Reactor-grade plutonium makes inefficient but dirty bombs. It also has a curious physical property of igniting spontaneously when exposed to air, thereby producing tiny aerosolised particles which are dispersed by wind currents and available for inhalation by humans and animals. One could envisage disastrous consequences if a truck were to crash and discharge some of its deadly contents. Plutonium must be transported very carefully, packed in small quantities in separate containers because only 10lbs is 'critical mass', which means that a spontaneous atomic explosion could occur if 10lbs or more were compacted together in a finite space.

The most awesome property of plutonium is its half-life of 24,400 years, (half-life of a radioactive substance being the period of time for half of a given quantity to decay, and a similar period for half of the remaining radioactivity to decay, *ad infinitum*.) Therefore radiation from man-made plutonium will exist on earth for at least half a million years. To illustrate the enormous medical problems arising from the physical properties of plutonium, if an individual dies of lung cancer engendered by plutonium, his body will return to dust but the plutonium lives on to produce cancer in another human being. Although it will be used as 'fuel' in breeder reactors, more plutonium will be produced than will be utilised. So there will be a continual net increase in plutonium manufactured. The nuclear industry has not yet decided what to do with all this plutonium; there are no safe methods of disposal and storage available at this point of time.

Waste storage. After the plutonium is extracted from the radioactive waste, biologically dangerous elements remain, which have no further use and are pure waste products. This remaining solution contains some plutonium, radioactive iodine, strontium-90, cesium and many other highly toxic radio-nuclides. Because it is extremely hot, it must be stored in tanks which are cooled continuously for years. Every month numerous leaks of radioactive wastes are reported in the United States in quantities from several gallons to 200,000 gallons. When this dangerous fluid leaks it will inevitably contaminate the water system of the planet, and the various elements will be taken up by the food cycle. Radioactive iodine, strontium-90 and cesium are absorbed by roots of grass and vegetables and are further concentrated in the flesh and milk of animals when they eat the grass.

Iodine-131, strontium-90 and plutonium are concentrated in both human and animal milk. Cesium is concentrated in muscle (meat) and plutonium is also a thousand times more concentrated in fish compared to the background water concentration. These substances are invisible, and because they are tasteless and odorless it is impossible to know when one is eating or drinking or inhaling radioactive elements.

Biological properties of radioactive waste. Genes are changed by radioactive particles. Cells and genes which are actively dividing (as in foetuses, babies and young children) are most susceptible to the effects of radiation. If a gene which controls the rate of cell division is altered by radiation, the cell may divide in an uncontrolled fashion to produce cancer and leukaemia. It may take from 15-30 years before the cancer appears after the cell is exposed to radiation. If a gene in the sperm or egg is altered by a radioactive particle, the young may be born either with an inherited disease, or the baby may appear normal but will transmit the damaged gene to future generations, to become manifest in later years.

Radioactive iodine is absorbed through the bowel wall, and migrates in the blood to the thyroid gland, where it may produce thyroid cancer.

Strontium-90 is also absorbed through the bowel after being ingested in contaminated milk, and is incorporated in bone because it chemically resembles calcium. This element causes osteogenic sarcoma — a highly malignant, lethal bone tumour — and leukaemia. The blood cells formed in the bone marrow are subjected to the effects of radiation from strontium-90 in the adjacent bone.

Cesium-137 is deposited in muscles of the body, where it can produce malignant changes.

Plutonium is one of the most carcinogenic substances known. It is not absorbed through the bowel wall, except in infants in the first four weeks of life when it is ingested in milk. As previously described, infants are extremely sensitive to the toxic effects of radiation. The route of entry of plutonium is by inhalation of contaminated air into the lungs. Small particles of plutonium are deposited deep in the respiratory passages, where they tend to remain for years. It is accepted that one millionth of one gram of plutonium is sufficient to produce lung cancer 15-20 years after initial inhalation of the element. Plutonium is also absorbed from the lungs into the bloodstream where it is carried to the liver (where, like strontium-90, it causes osteogenic sarcoma and leukaemia), and it is selectively taken up from the circulation by the testes and ovaries where, because of its incredible gene-changing properties, it may cause an increased incidence of deformed and diseased babies, both now and in future generations.

Plutonium also crosses the placenta, from the mother's blood into the blood of the foetus, where it may kill a cell responsible for development of part of an organ, e.g. heart, brain etc., causing gross deformities to occur in the developing foetus. Production of foetal deformities is different from the deformities caused by genetic mutation in the egg or sperm, because although the basic gene structure of the cell of the foetus is normal, an important cell in the developing foetus has been killed leading to a localised deformity. (Similar to the action of the drug thalidomide).

Massive quantities of radioactive wastes are being and will be produced in the future. The safe storage of waste is unsolved, and even if there were a present-day solution, we could not predict a stable society or world for half a million years. We could not guarantee incorruptible guards or moral politicians and we certainly cannot prevent earthquakes, cyclones or even wars. As waste is leaking now, so inevitably will it leak in the future. We could therefore predict epidemics of cancer and leukaemia in children and young adults, and an increased incidence of inherited disease (there are 2,000 described inherited diseases). It is also inevitable that plutonium will be stolen and utilised for atomic weapon production; two tons of plutonium are presently unaccounted for in the United States.

It has been claimed that 80-90 per cent of all cancers may be caused by environmental pollutants. There was a 5 per cent increase in cancer in the United States in the first seven months of 1975, and a total 3 per cent rise in 1975.

Governments spend millions of dollars researching the causes of cancer, leukaemia and inherited disease, but simultaneously spend billions of dollars in an industry that will directly propagate these diseases. As a doctor, I appeal to my fellow medical colleagues to investigate this enormous present and potential threat to our patients.

Dr Caldicott, who was the 1960 winner of the BMA prize for clinical medicine, undertook a European lecture tour in the autumn of 1980, during which she spoke to doctors at University College, London. *Vole* magazine (November 1980) reported that the audience included Nobel prize-winner Maurice Wilkins, of the British Society for Social Responsibility in Science, and that a working group is now planning to set up a British equivalent of Physicians for Social Responsibility. The need for such a movement is becoming increasingly obvious. The *Daily Telegraph* (15 November 1980) published a letter headed 'Atom Danger to Mothers', written by a former Harwell worker, Margaret Bullock, who said:

I was interested in your recent report that the Medical Research Council was proposing to screen all former exployees of the Atomic Energy Authority. In January 1957 I started work at the UKAEA,

Harwell, and remained there for five years. During this time I handled a quantity of radioactive material while working at one of the reactors. I left to get married and during the next nine years had a disastrous gynaecological history. I have had five pregnancies, and have one surviving child who is severely mentally handicapped. My room-mate in the staff hostel also had a severely handicapped child who, I believe, is now in a home. Doctors over the years have pooh-poohed any suggestion that my work experience could have had any effect on my inability to bear a 'normal' child. If the Medical Research Council is to do this study I suggest it uses the opportunity to research the incidence of abnomal births in former employees.

At the Guy's Hospital Community Health Conference on Low-Level Ionising Radiation, at which Dr Morgan spoke, Dr Rosalie Bertell, professor of mathematics at New York State University and statistician for the Roswell Park Memorial Institute, described the findings of the huge Tri-State study carried out in the US. In this study she discovered a clear correlation between leukaemia and levels of radiation exposure well within the guidelines proposed for nuclear workers and the general public by the International Commission on Radiobiological Protection (ICRP). In the light of such claims by highly reputable scientists and doctors that there is no safe or 'permissible' exposure to radiation, it is instructive to observe the levels of radioactivity being routinely dispersed into our environment by the nuclear industry. In *Nature* (vol. 284, March 1980) the Secretary of State for the Environment gave the following figures for the amounts of radioactive wastes produced *each year* by the UK civil and military nuclear programmes:

$100m^3$ of high-level liquid wastes
$500m^3$ high-level solid wastes
$450m^3$ of plutonium-contaminated wastes
$250m^3$ of miscellaneous wastes

Plutonium-contaminated waste, and medium-level and low-level wastes are dumped at sea, as is waste from the decommissioning of reactors (a programme which is only just beginning). The total amount of civil waste in store at the end of 1979 was approximately $20,000m^3$, including:

$1,000m^3$ at Windscale — liquid
$700m^3$ at Dounreay — liquid
also $9,000m^3$ of high-level solid, including $3,500m^3$ plutonium-contaminated waste.

Cesium-137 discharged into the sea at Windscale from the waste
reprocessing plant has been discovered off the coast of Ireland,
and in fish in the Baltic Sea, although there was a two-year delay
in reporting this fact from the Government Radiobiological
Laboratories at Lowestoft. The 1978 report on radioactive emissions to the environment, issued by the Department of the
Environment at the request of the Royal Commission on Environmental Pollution headed by Sir Brian Flowers, noted a
'marked increase in plutonium-241 discharges into the sea'. It
reported:

11 tonnes of uranium discharged into the sea
Strontium-90 discharge rose from 11,534 curies in 1977
(34 per cent of limit) to 16,160 curies in 1978 (54 per
cent of limit)
Plutonium-241 discharge rose from 26,517 curies in
1977 to 47,928 curies in 1978

Although the report describes plutonium-241 as of 'low-energy
and low radio-toxicity', it decays into a smaller quantity of the
highly-toxic radioisotope americium-241. The build-up of
americium is now a major issue, and new limits for
plutonium-241 are currently being defined. Gaseous discharges
into the atmosphere consisted of:

Krypton-85   — 700,000 curies in 1978
Argon-41   — 29,000 curies in 1978
Tritium   —   6,000 curies in 1978

The argon release is considered to have exposed the local
population to 2.5 per cent of the international dose limit. The
calculated maximum dose to local fishermen was 30 per cent of
the 1977 international limit. However, as the Royal Commission
on Environmental Pollution observed in its *Sixth Report: On
Nuclear Power*, in 1976:

Nearly all the plutonium that is currently discharged to sea at

Windscale ends up on the bottom sediments. Some of them move into the Ravenglass Estuary. They are thought to be radiologically insignificant at present, but it is known that the estuary contours are changing with time, because of the net landward movement of sediment. Thus in about a century our descendants may be faced with two new exposure pathways for this plutonium. Sediments that are above high tide level and are no longer wetted may blow about in dry, windy weather and possibly form respirable aerosols. A smaller hazard could arise from grass growing on newly-reclaimed land, which may support cattle, as happens now. They would require monitoring ... This is a serious issue for the future.

It appears even more serious when one learns that the limit, or 'acceptable body dose', of plutonium has been set at less than one millionth of a gram, an invisible particle. Dr Caldicott notes that there is evidence that even this level has been set too high.

On 12 September, 1979, the *Guardian* reported that British Nuclear Fuels Limited had admitted to dumping 1,500 gallons of radioactive waste each week into the sea near Liverpool. This fluid, containing technetium-99 and uranium, comes from the uranium enrichment plant at Capenhurst which prepares fuel for use in nuclear reactors. A recent BNFL report claims, however, that this dumping is 'completely safe and will have no effect on holiday bathers or fishing'. Yet the Irish Sea is known to be the most radioactive sea in the world: radiation levels have increased ten-fold in recent years. The Channel off the Contentin Peninsula in Normandy is polluted within a radius of 100km with radioactive waste from the nuclear installations at Cap de la Hague near Cherbourg.

It is instructive to look at the statistical pictures of cancer or cancer-related deaths which are now being observed near large nuclear operations. Dr Robert Blackith, of the Department of Zoology, Trinity College, Dublin, compiled the mortality rates published by Cherbourg (including la Hague and Cotreville with their large nuclear generating and reprocessing plants), *excluding* cases of leukaemia and patients who happened to be taken to the hospital at Caen:

	1971	1979	percentage increase
cancer	74	197	165%
heart disease	86	103	20%
cardio-vascular	66	80	20%
liver cirrhosis	26	36	35%

The Institute of Biological Safety at Bremen has calculated that the rate of leukaemia deaths in children under 15 years old near the Lingen reactor is now six times greater than before the installation of the reactor in 1968. Dr Gerald Drake has recorded a significant excess of leukaemia and brain tumours near the Big Rock Point nuclear plant in Michigan during the years 1972-75; and in 18 counties bordering an area containing three boiling-water reactors, the rate of leukaemia, despite increased public health care, is declining less rapidly than elsewhere in the country. Increases in leukaemia rates have also been found in Anderson City, near the Oak Ridge uranium enrichment plant. A study of increased cancer incidence downwind from the US Rocky Flats nuclear plant, by Carl Johnson, was presented at the International Radiological Protection Agency Congress in Jerusalem, March 1980. In Australia Dr Bruce Brown has associated the radiation fallout rate, even where significantly lower (2-3 millirems) than the 'permissible' limit, with a sharp rise in the leukaemia rate among males — many of whom work outside in dry, dusty, windy areas.

In our country, Dr Geary of the University of Manchester has recorded an outbreak of myeloid leukaemia in North Lancashire — over the last ten years the incidence rate has nearly doubled. North Lancashire is an area covered by the prevailing wind from Windscale, and is also an area where a great deal of locally-caught fish is eaten. The town which has the highest incidence of leukaemia in England is Barrow-in-Furness, southeast of Windscale where the prevailing wind is from the northwest (this information was supplied in a written reply to a parliamentary question submitted by Mr Pavitt, MP, Labour, Brent South : *Hansard*, 26 October 1979). In September 1980, balloons were released from the proposed site for a nuclear reactor at Langton Herring in Dorset. The balloons were labelled: 'I am a radioactive particle routinely emitted from the proposed nuclear power station on Chesil Beach in Dorset. Please return me if you find me (address given); we must all know how far I can reach.' Balloons were returned from Belgium, Germany and the Isle of Wight.

On 30 November 1979, the *Guardian* reported that child leukaemia is double the normal rate among the population of Utah and Nevada exposed to low-level radiation from the 1953 test blasts. *The Times* of 2 April 1980 carried the following report under the headline 'Big rise in baby deaths near nuclear power plant':

> Deaths of babies below the age of one doubled within a ten-mile radius of the Three Mile Island nuclear power station in the six months after the accident there a year ago. The figures were revealed by Dr Gordon Macleod, who was Secretary for Health in Pennsylvania when the accident occurred. There were 31 such infant deaths between April and September 1979, compared with only 14 in the same period in 1978. Other figures have shown thyroid abnormalities in babies born near the site since the accident to be one in 925, compared with a national average of one in 5,000 ... Dr Macleod was dismissed from his state post last September and is now a professor at the University of Pittsburgh. The figures are sure to increase the apprehensions of people who live near the plant about plans to vent radioactive krypton gas still trapped in it. The owners of the power station want to release it into the atmosphere slowly over two months, but the protesters insist that a safer way must be found. The residents argue that they cannot believe official protestations that there is no danger, bearing in mind the misleading information from the power company during the accident. One couple is already suing the owners and designers of the plant for damages, claiming that their baby was stillborn because of the radioactivity released during the accident.

Controversial studies have been made on workers in the huge fuel reprocessing plant at Hanford, Washington and the nuclear shipyards at Portsmouth, New Hampshire — controversial because the US Navy withheld information on its Portsmouth workers from investigators, and because funds were withdrawn from the scientists studying the Hanford workers before the study had been completed. Funds were also withdrawn from the Tri-State investigation being carried out by Dr Bertoll and scientists from the University of Minnesota, and the Johns Hopkins

University of Baltimore — but only when this statistical study of leukaemia rates began to show a correlation with low-level radiation exposure. The Atomic Bomb Casualty Commission, set up in 1949-50 to study the after-effects of the bombing of Hiroshima and Nagasaki (on those who survived) will not release the data it has gathered, although such data provide a principal source for the 'safe' standards set by the International Commission for Radiological Protection. It should also be remembered that the people who survived the five years that elapsed between the bombing and the setting up of the Commission are, by definition, a 'survivor' population and therefore stronger than average. Dr Dolphin of the National Radiological Protection Board has pointed out that radiation exposure levels in this country are set by the average population, not by sensitive sub-groups such as children or pregnant mothers. Workers in nuclear power or processing plants are chosen for their good health record and comparative youth; yet even here distressing deaths are occurring and liability being questioned in court cases.

A particularly hotly debated case is that of a 49-year-old Aldermaston worker, Mr Cummins, who died of a rare form of cancer two years after being exposed to an excess of the 'permitted' radiation dose. The foreman said that Cummins had obeyed all safety regulations very carefully; Dr Mole, of the Medical Council and the National Radiological Protection Board (NRPB), claimed that 'natural background radiation' was a greater source of danger to Mr Cummins than overexposure at work, and that it would take far longer than two years for such a rare cancer to develop. However, Professor Patricia Lindop (who holds the Chair of Radiation Biology at St Bartholomew's Hospital in London) disagreed, feeling that such a cancer could have no other clear cause, and that it was much more likely to be radiation-induced that to occur naturally. What is particularly disturbing about this death is that the British Nuclear Fuels Limited doctor at Windscale, Dr Schofield, said that an autopsy revealed that Cummins had only 1 per cent of the 'permitted body burden' of radioactive material (*Guardian*, 30 November 1979).

Nor is cancer the only fatal illness connected with radiation exposure. Another nuclear worker from Aldermaston, Mr Higgins, inhaled 300 times the permitted dose of ruthenium-106 in 1973, with resultant damage to liver and lungs. He was later found to have suffered serious damage to his thyroid and parathyroid glands, symptomatic of serious exposure to radioactive iodines, which suggests a separate, unpublished accident. The combination of damage to liver, kidneys and thyroid can cause the body's failure to retain potassium in muscle tissue, and such abnormal muscular effects can bring about subsequent heart failure — of which Higgins died. Such cases, as well as the case of the Windscale worker Malcolm Pattinson, whose widow was awarded £67,000 damages by BFNL when he died of leukaemia after eight years' work in radiation areas, suggest that the CEGB's use of *averages* in measuring and setting limits for workers exposure is worse than inadequate. Another Aldermaston worker, Mr Ansell, died on 4 October 1980, aged 60. He had already filed a claim (still outstanding) against the Ministry of Defence for radiation injury. He was cremated without a post-mortem. In fact, he died the day before the re-opening of the adjourned inquest on his fellow-worker, Ronald Kent. Professor Lindop alerted other claimants to the importance of postponing cremation and preserving medical evidence. Ansell died of cirrhosis of the liver and peritonitis, both known characteristics of radiation damage as demonstrated by animal experiments involving plutonium and other fission products. The coroner suggested that Ansell might have been a drinker, a fact which was hotly denied by his family and friends. His wife said: 'One time he came home frightened out of his life. He had had to go back again and again to the monitoring machine because he couldn't get the stuff off his hands. I know he had to work in special clothing a lot of the time, and I think he kept quiet so as not to frighten me.' Two years previously, Ansell had complained of gastric ulceration and weight loss; a fact which leads one to ask if Ansell's doctor was aware that Ansell was a radiation worker (*Guardian*, 21 October 1980).

The fact that BNFL withheld some of the details of Pattinson's case from the public, and, with the Atomic Energy Authority, settled other widows' suits (Connors, Troughton) out of court, does not inspire confidence in their claims of having established 'safe limits'. Furthermore, the statement in the NRPB's report published before the Windscale Inquiry, that there was no statistical excess of cancer among Windscale workers, subsequently had to be retracted during the inquiry — Dr Dolphin of the NRPB admitted that the significance levels had been incorrectly calculated. What is more frightening to the public in general is that Dr Dolphin had been informed of this discrepancy by the NRPB statistician, Dr Garson, *before* he published the report. Is it only coincidence that on 6 April 1978, after the Windscale Inquiry, the NRPB advised the government that the safe limits for whole-body radiation be reduced five-fold overnight? It would be interesting to know how the new levels were explained to the fishermen and those who ate the fish they caught in the Irish Sea, who had thus been exposed to radiation levels 2.5 times the new dose limit.

Calculating radiation exposure for workers by average exposures cannot be considered satisfactory. On paper, it appears that the more workers there are, the less radiation each will absorb. In reality, each worker is exposed to the full excess dose in that space, and no low dose can offset or undo a high dose. Yet this is how the NRPB dismisses the difficult problem of worker exposure. According to a report published in France in 1976, out of eight cases of leukaemia among workers at Cap de la Hague, three of the workers were aged between 40 and 50 and were classed as 'occupational deaths'. The risk of spina bifida in children born to reprocessing workers in France appears to be ten times the national average. Yet the French workers' exposure is roughly half that at Windscale. Perhaps that is one reason why the British nuclear industry has not introduced a health follow-up system on workers who have left the industry, or on their children.

A further extremely dubious use of the 'average dose' calculation is in the 'burning out' or 'sponge' technique. This is used when radiation levels are so high, in an area where vital work must be carried out, that the industry allows the workers concerned to receive their total yearly permissible dose in the space of one or two days, or even a few hours. Since the work in question is usually specialised, this is as inefficient as it is dangerous to the workers. Dr Karl Morgan refers to the highly questionable practice, common in the US, of hiring temporary workers off the streets, at very high wages, to carry out such high-radiation repair or maintenance. Added to the risks which are inherent in such high-level exposure, there is the additional problem that casual workers are often inadequately trained for the work that they have to carry out.

Not only is this practice morally and technically suspect, there is a also a genetic risk. Dr Peterson, who served on the Kemeny Commission that investigated the Three Mile Island accident, pointed out in a speech to the American Association for the Advancement of Science the threat to the population in spreading incalculable amounts of radiation genetically by the use of humans as radiation 'sponges'. And yet, because of the enormity of the hazards and the critical nature of the nuclear processes, such methods are necessary for the continued day-to-day running of the nuclear industry — for we have not been referring to work arising from accidents or explosions, but to the normal operating procedures of nuclear installations. The horrifying suffering and death of an American nuclear worker, Joe Harding, may illustrate Dr Peterson's point. After working for 18 years in a uranium enrichment plant in Kentucky, Joe Harding died of a combination of radiation-induced cancers and a hitherto unknown form of pneumonia. Before he died, 'in 1970, nail-like growths began developing from the finger-print side of Joe's fingers and thumbs. Somewhat later they started growing also from his knuckles and finger joints. "Now I have fingernails growing even from my wrists, elbows and shoulders," Joe says. "And something like toenails are growing from my ankles and kneecaps. Various doctors have said it is mutations, cell changes caused by radiation." ' (*Ecologist*, 10 December 1980)

> 'It is possible that the degree of radiological safety we demand is incompatible with the existence of a viable nuclear industry. If this is true, we must be prepared to accept some fundamental changes in our society.' — Dr Irving Lerch, New York Medical Center. (*New Scientist*, 3 January 1980.)

At the Guy's Hospital conference, Dr Bertell described the International Commission on Radiological Protection, formed in London in 1950, in these words:

The ICRP, together with the International Commission on Radiological Units, seems to have acquired an undeserved reputation as an independent scientific advisory body on worker and general public exposure to ionising radiation ... but membership of the ICRP is contingent upon nomination by national delegations to the International Congress of Radiology and by ICRP members, subject to approval and selection by the ICRP International Executive Committee. It is a self-perpetuating club, not a scientific society based on professional excellence. ICRP recommendations for human exposure to ionising radiation and ICRP estimates of the probable effects of such recommendations have become increasingly suspect over the years. The latest publication, ICRP 26, contains a most blatant violation of rational scientific procedure. It states that ICRP predictions of health effects deserve absolute credence and take precedence over all observations of fact. Paragraph 184 contains the following assertion: 'Medical surveillance has no part to play in confirming the effectiveness of a radiation programme.' This amounts to a claim that no audit of human health can ever challenge the ICRP forecast. This is especially unscientific in the face of exposure of biologically diverse people with hereditary, environmental and host-defence mechanisms varying in unknown ways.

Dr Bertell also points out the irresponsibility of the claim made by the nuclear authorities that the low-level radiation poured into our environment by the industry is less than natural radiation and therefore must necessarily be harmful. Such an assertion fails to take into account the fact that natural background radiation is radio-potassium, which does not concentrate in any specific part of the body, unlike man-made radio-nuclides. Biologist Professor Ichikawa said on BBC Radio 4 on 24 April 1980: 'Man-made radio-nuclides can be concentrated very highly in man — this is a very recent discovery, which would lead me to veto nuclear power.'

World Medicine (20 September 1980) reports:

Dr Paul Nickson of Bedford found the following paragraph in *Atom News*, produced by the information (*sic*) services branch of the UKAEA: 'In Britain radio-isotopes are often used to help predict a difficult birth. Over a million tests a year are done in Britain, where 96 pregnancies out of every 100 result in the birth of normal healthy children.' If, asks Dr Nickson, the UKAEA can publish such misleading rubbish as 'information', how can people who know little about nuclear energy 'believe anything this government-funded propaganda machine churns out?'

Perhaps even the CEGB are beginning to understand the unscientific and even inhuman nature of their claims, however irrationally this understanding may be expressed. In February 1979, in the radiation section of *Nuclear Engineering International*, Paul Woolham of the CEGB argued that the nuclear industry could not be accused of causing any *extra* deaths — although those deaths perhaps happened *earlier* than they would otherwise have done. BNFL make an equally amazing claim in their expensively-produced brochure, 'Atoms for Energy': 'By removing background radiation (in the form of uranium) from the environment we are making it safer for everyone.' Short of risking a suit for libel, it is difficult to know what comment to make on such statements.

2. Wastes

The 'routine' emissions of radiation described in the previous chapter do not include the equally 'routine' production of high-activity waste for which no one has as yet found any safe means of disposal. The report of the Royal Commission on Environmental Pollution issued in 1976 (the Flowers Report, after the commission's chairman, Sir Brian Flowers) commented: 'We must assume that these wastes will remain dangerous and will need to be isolated from the biosphere for hundreds of thousands of years. In considering arrangements for dealing safely with such wastes, man is faced with time scales that transcend his experience.'

The same report also notes, rather more stringently, in paragraphs 391 and 427:

Neither the Atomic Energy Authority nor British Nuclear Fuels Limited in their submissions to us gave any indication that they regarded the search for a means of final disposal of highly-active waste as at all pressing, and it appears they have only recently taken firm steps towards seeking solutions. We think that quite inadequate attention has been given to this matter, and we find this the more surprising in view of the large nuclear programmes that both bodies envisage for the coming decades, which would give rise to much greater quantities of waste ... The picture that emerges from our review of radioactive waste management is in many ways a disquieting one, indicating insufficient appreciation of long-term requirements either by government departments or by other organisations concerned. In view of the long lead times that will almost certainly be involved in the development of appropriate disposal facilities, we are convinced that a much more urgent approach is needed, and that responsibilities for devising policy and for executing it need to be more clearly assigned.

Since 1977, well after the publication of the Flowers Report, the National Radiological Protection Board, at the request of BNFL, has spent £70,000 on assessing the radiological consequences of the disposal of radioactive waste in geological formations. But in comparison with the amounts of public money devoted — without public consent — to the construction of ineffectual new nuclear power stations, the sum spent on all-important safety research is derisory.

In 1976 the exiled Soviet scientist Dr Zhores Medvedev, who now works for the Medical Research Council in London, reported the vast scale of an accident, apparently nuclear, which occurred in the Urals (USSR) in 1957. Medvedev's report led scientists at America's main nuclear laboratories at Oak Ridge, Tennessee, to study Soviet maps of the Urals before 1958 and after 1970. Their study confirms that a huge release of radioactivity apparently occurred in this area. Small communities and large towns have been removed from the maps. Although this is a well-watered region, which would normally have no need for canals or reservoirs, the natural river and lake system has been altered by a new canal system which apparently bypasses the most contaminated area. One large lake (Kzyltash) is now isolated from the river system. Medvedev theorised an explosion of radioactive waste buried underground, without yet knowing how such a powerful explosion could have come about. The Oak Ridge laboratories have now proposed a possible explanation — that the ammonium nitrate formed in the wastes may have exploded, with the possible force equivalent to 100 tons of TNT.

Such an explosion may have ruptured another waste storage tank nearby, which contained nuclear waste of a different composition. This could then account for the pattern of contamination deduced from Soviet maps of the area, the recorded population increase in adjacent towns, and the appearance of unusual papers in Soviet journals dealing with biology, genetics and botany.

Moreover, other more immediately dangerous problems are still apparently considered unimportant. Nuclear waste has to be transported from power stations throughout the country to the reprocessing plant at Windscale. There have already been three derailments of trains carrying nuclear waste in England; the flasks containing the high-activity waste have not been tested for impacts over 30mph, or for fires lasting longer than 30 minutes. Two of the regular routes pass through London and if such a high-activity flask should break, an area covering a radius of three miles will be uninhabitable for 70 years, and there will be heavy contamination of an area within a radius of 12 miles. Once the waste has reached the reprocessing plant, the dangers are by no means over, as Peter Bunyard points out clearly in *The Ecologist*, April-May 1980:

> Reprocessing to extract weapon-grade plutonium presents no fundamental technical problems, although it creates high-level waste which has to be kept in stainless steel tanks with continual refrigeration and stirring to prevent heat build-up and the settling out of solids. Similar reprocessing techniques have been developed for Magnox fuel with its burn-up of up to 5,000 megawatt-days per tonne, again with the production of high-level waste and of medium and low-level waste, some of which is discharged into the environment. Reprocessing of Magnox fuel, and the French equivalent UNGG fuel, has been carried out on an industrial scale at Windscale in Cumbria, and at Cap de la Hague in the Cotentin Peninsula of Northern France. Corrosion of the magnesium alloy cladding when the spent fuel is put in cooling ponds has given BNFL some headaches at Windscale, primarily because of the escape of cesium-137.
>
> In the early 1970s BNFL had to ask for authorisation to increase its discharges, particularly of beta-emitters, and in 1977 alone it discharged 121,000 curies of cesium-137 out of a total beta-emitter discharge of more than 192,000 curies. On account of the corrosion problem, reprocessing of the spent Magnox fuel is con-

sidered mandatory, and both Britain and France have gained considerable experience on this fuel alone. As pointed out, the reprocessing technology as applied to Magnox fuel is considerably simpler than that needed for higher burn-up fuel. Yet Magnox fuel reprocessing has hardly had a satisfactory record. The industrial plants have aged very rapidly, giving no more than 10 to 15 years of service before needing to be replaced. The discharge from the reprocessing plants, into the atmosphere and the sea, is already close to the authorised limits, themselves raised, and the workers in the plants have been subjected to relatively high radiation doses which come in the range of those found to enhance the risk of cancer.

The history of reprocessing hardly bodes well for the future when high burn-up fuels are used. Much of the poor publicity in Europe concerning reprocessing plants has been directed against Cogema, the operators of Cap de la Hague, who have been accused of subjecting the reprocessing workers to awful working conditions on account of radiation leaks. Indeed, at the end of 1976 the reprocessing workers at Cap de la Hague went on strike for better conditions. But a simple comparison of the two reprocessing plants, using official figures, shows Windscale in an even worse light than Cap de la Hague. Under its authorisation Windscale is permitted to discharge 300,000 curies of beta-emitters each year and up to 6,000 curies of alpha-emitters. Cap de la Hague is permitted to discharge 45,000 beta and only 90 curies of alpha. In 1978 Windscale discharged 64 per cent of its beta allowance and 31 per cent of its alpha. Since Cap de la Hague has a throughput of one-third that of Windscale, the latter on a weight for weight basis is actually discharging 1.5 times more beta than the French plant is authorised to discharge, and 7.5 times more alpha. In 1978 Windscale discharged more than 10 tonnes of uranium, and some 48,000 curies of plutonium-241, itself a beta-emitter, which, with its half-life of 14 years, then decays into the longer-lived alpha-emitter americium-241. Like other alpha-emitters, americium gradually builds up in the environment; like them too it is extremely radiotoxic (when breathed in, as from drying mud flats, or eaten, as in sea food from the Irish Sea in particular).

In addition to these conditions, unquestioningly accepted as permanent features by the nuclear authorities, there are also unexpected and unaccounted-for accidental emissions of radioactive waste, such as that from the fire at Windscale in 1957, which permanently closed part of the plant. Some 20,000 gallons have leaked from the high-activity water-cooled silos at Windscale; this leak is still uncontrolled and is spreading into the ground and, ultimately, into the water-table. Plutonium is kept under maximum-security guard — though 16kg is officially 'unaccounted for', missing from Windscale alone in the one year 1976-77. (Six to eight kg will suffice to make a bomb, less if you are clever). Of the high-activity wastes which are dissolved in acids and stored for a limited period in stainless steel tanks, the Flowers Report said: 'these liquors are not particularly large in terms of volume, but their radioactive contents, and therefore their aggregate toxicity, are immense.' The CEGB, in their 1979 glossy pamplet, *The CEGB and Nuclear Power: Questions and Answers* confront the question of such wastes thus: 'All the highly-active liquid waste so far produced as a result of Britain's nuclear power programme would scarcely fill a four-bedroomed house.'

There is still no known safe and permanent method for disposing of this waste. Researchers are trying to perfect a method called vitrification, which involves incorporating the waste into glass blocks, which could be more easily, but perhaps irretrievably, dumped. However, in the *Observer* (28 October 1979), Jeremy Bugler reported that scientists at the University of Pennsylvania had discovered that these blocks containing waste broke up under high water pressures and high temperatures. As it had been hoped to bury them deep underground or in the ocean, the local pressure ranges would be considerable; and radioactive waste generates so much heat that underground storage would automatically create high temperatures. The news from Pennsylvania appears not to have reached our present Under-Secretary for Energy, Norman Lamont. Speaking in the House of Commons, a month after the publication of Jeremy Bugler's article, he said that he was satisfied that vitrification proposals and present plans for storing waste work safely. The Pennsylvania research has been confirmed by recent work in Australia — the *Guardian* (8 October 1980), reported that scientists at the Australian Defence Research Centre in South Australia have found that radiation renders vitreous materials — such as the boro-silicate glass proposed for nuclear waste disposal — highly susceptible to chemical attack and breakdown by moisture. It was hoped that the process of vitrification would facilitate disposal of high-level waste — but even had it been found to work technically, it is important to remember that vitrification would not have stopped the radiation, merely made the waste easier to handle.

Even if the vitrification process had been conquered, the question remains of where such lethal glassware should be deposited. Galloway? Somerset? Our already crippled oceans? Public opposition to drilling applications has been vociferous — in the Cheviots the whole range of public opinion was mobilised against the application at a public inquiry, and in Somerset the parish council at Puriton passed a unanimous resolution to ask MP Tom King to ensure that drilling is not carried out. There can be no doubt that radioactive particles would seep out into the water-table or the sea-bed, even assuming that the deposit areas are geologically stable for the required half-million years or so; if this assumption proves incorrect (one is reminded of the several recent earth tremors near Windscale) then radioactivity would be released directly into the environment. The consequent environmental hazards would be innumerable.

3. Accidents

The ultimate environmental catastrophe directly caused by the process of nuclear power generation is, of course, an accident originating in a malfunction of the power station itself. Such an accident may be the result of human error, a failure in the technology or in the materials by which we try to control such tremendous power, or some characteristic of the nuclear fission process of which we are as yet unaware. The long and horrifying catalogue of accidents in nuclear power stations in many countries demonstrates how all those factors appear to have contributed at one time or another. The potential environmental effects are so gross that they have never yet been adequately presented to the public. Richard E. Webb is a former nuclear reactor engineer who left the industry 'to study full-time and without constraints the accident hazards of nuclear reactors in all essential aspects'. He sees the problem as a combination of scientific and social difficulties:

In order to assess the risks of reactor accidents, therefore, it is necessary to examine the reactor explosion potential and the associated scientific uncertainties; and to question the practicality of an adequate research safety programme to resolve these uncertainties. Furthermore it is necessary to assess the likelihood of serious accidents. For this task we need to appreciate the many different ways accidents can occur and the many subjective judgements inherent in any evaluation of reactor safety, and to analyse the experience of actual reactor mishaps and malfunctions. Also, in order to assess the hazards, we will need to carefully analyse in detail the possible harmful consequences of a given heavy release of radioactivity to the atmosphere, including a social determination of what constitutes 'an acceptable emergency dose' of radiation to the populace, an important part of the 'WASH-740' report (an AEC report on possible nuclear accidents, commissioned in 1957). We must try and foresee all possible effects and complications, such as the potential for atmospheric down-draughts that might swoop a radioactive cloud down on to a metropolitan area; the concentration of radioactivity in rainwater in roadside drainage ditches running along evacuation routes, which might conceivably cause lethal doses of radiation to a fleeing population; the potential for panic in an evacuation; the public health burden on communities which must accommodate large numbers of refugees from contaminated areas; the potential for contamination of sources of drinking water, such as the Great Lakes; and the need for both shielding and the closing of windows, doors and other openings to avoid high radiation exposures when taking shelter in advance of an approaching radiation cloud.

Richard Webb wrote these words in 1976. There have been several theoretical accident reports on nuclear safety, the first being the WASH-740 report which he quotes. He has updated this report, to account for the six-fold increase in the highly-intense, short-lived radioactivity and the fifteenfold increase in the long-lived radioactivity in contemporary reactors (notably the pressurised water reactor which the present government hopes to install throughout Great Britain). He has also taken account of the increased fuel temperatures achieved by the use of uranium oxide, rather than by uranium metal which was the fuel material in use at the time of the WASH-740 report. The oxide fuel melts at a much higher temperature, which means a greater chance of boiling off radioactive substances, a greater fractional release of radioactivity from the reactor, a finer radioactive dust, and a stronger explosion for expelling the dust. He calculates the results in these terms:

The maximum conceivable consequences of the worst accident are as follows:
(1)   a lethal cloud of radiation with a range of 75 miles and a width of one mile;
(2)   evacuation or severe living restrictions for 120,000 square miles, lasting a year or possibly longer; and
(3)   severe long-term restrictions on agriculture due to strontium-90 fallout over 500,000 square miles, lasting one to several years, with dairying prohibited 'for a very long time' over a 150,000 square mile area.

There are other consequences not here estimated for water-cooled reactors, such as genetic damage. The potential accident consequences for the fast-breeder reactor — especially with respect to plutonium contamination, which may be a gravely serious lung cancer hazard ... will depend on the explosion hazard unique to that reactor. Incidentally, the maximum distance downwind from a reactor accident associated with the above land-area estimates of severe living restrictions and agricultural restrictions is about 1,500 to 2,000 miles. Hence, a nuclear reactor accident could affect distant communities as well as those nearby ...

Moreover, the WASH-740 estimates were not meant to be upper limits, as worse weather conditions for promoting damages were not included: 'This study does not set an upper limit for potential damages; there is no known way at present to do this.' Hence, the above estimates for the maximum accident consequences for present-day reactors, based on extrapolating the WASH-740 estimates, are not really upper limits either. The above extrapolated levels of disaster will depend on whether a large fraction (50 per cent) of the radioactivity can be released from the reactor into the atmosphere upon an accident; and released in the form of a very fine, light dust, namely, dust particles one micron diameter in size, so that it can disperse over a wide area before falling (one micron is about one forty millionth of an inch). Whether such a release of radioactivity can occur will depend on the *reactor explosion potential*, which in turn will depend on the fuel temperature levels attainable in a reactor accident. This is because the radioactivity generated by the reactor builds up within the solid fuel material, so that in order for the radioactivity to escape the reactor, the fuel must overheat and melt or vaporise by a reactor accident. If the fuel melting temperature is high enough, most of the radioactivity would tend to boil, creating radioactivity vapour (smoke). Fuel melting then enables the radioactive vapours to bubble out of the fuel, and escape the reactor. Of course, if the fuel vaporises, which occurs at higher temperatures, the radioactivity would vaporise right along with it. Furthermore, such hot, molten or vaporised fuel could potentially cause an explosion, which could rupture the reactor enclosure and thus allow the escape of radioactivity (radioactive smoke) into the atmosphere.

This original report has been followed by several others — the Barber Report, the Brookhaven Report, the Lindackers Report, the Lewis Report and the Kendall-Moglewer Report. However, the only report quoted by the nuclear industry, and used as a safety standard by the AEC and the UKAEA as well as European nuclear combines, is the Rasmussen Report, published in 1974. This report is already famous — or infamous — for predicting that the chances against the accident at Three Mile Island were a million to one. It used data provided by the US Environmental Protection Agency. The director of that agency protested that the Rasmussen Report had abused his data, and underestimated the dangers by a factor of ten. The technique of assessing accident probabilities used by the Rasmussen Report has been abandoned by NASA because it has proved too low, from experience. Yet the National Radiation Protection Board and the UKAEA use this report as the accepted — or quoted — safety standard. Richard Webb has pointed out that the report is 'grossly inadequate in scope' and contains 'crucial and unsubstantiated assumptions'; many of the test situations used were well under the scale of reactors in commercial use, and two forms of worst possible accidents are not dealt with at all. Dr Webb pointed out these failings a year before the accident at Three Mile Island.

One of the most dramatic results of the American Freedom of Information Act has been the publication of the Nugget File. This document, a formerly secret catalogue of nuclear accidents and malfunctions spanning the previous ten years, was published by the Union of Concerned Scientists in 1979. In its margin, Dr Stephen Hanauer, a senior official with the Nuclear Regulatory Commission, pencilled the comment: 'Someday we will all wake up'. The dawning realisation of the potential consequences of their work explains the many resignations of safety officials and scientists within the nuclear industry. Among these are government scientists Dr Gofman and Dr Tamplin and the three nuclear engineers who resigned from GEC in New York because of the 'magnitude of the risks, the uncertainty of the human factor, and the genetic unknowns — we don't really know what goes on in a reactor — there should be no nuclear power'; as well as the USAEC nuclear safety engineer Carl Hooevar, who resigned because of the extent to which the public was being deceived about the consequences of 'minor' nuclear accidents.

A letter to the *Guardian* in August 1980, from Jean Emery of the Barrow and District Action Group Against the Import of Nuclear Waste, reveals extensive potential hazards which do not appear to be taken into consideration fully — if at all — by the nuclear authorities:

Barrow receives all the foreign spent nuclear fuel that comes into the United Kingdom; Magnox fuels from Europe and oxide from Japan. BNFL has argued that the oxide is fuel, not waste, and that it is being brought for reprocessing at Windscale. However, this has never been reprocessed successfully on a commercial basis anywhere in the world. In 1976 when BNFL did try to reprocess some of this material the result was that 35 men were contaminated and part of the plant had to be sealed off. So it remains that, until it can be used as fuel, it is waste. In the town itself, we have the nuclear submarine reactors which are in the same dock basin as the terminal for the Magnox and oxide ships. On the outskirts, a gas terminal is to be built, whose gas-condensate tanks are to be next to the proposed site for the docking of nuclear material. These gas tanks will be fed by a pipeline over which will pass nuclear-powered submarines, the nuclear carriers, and the gas-condensate carriers.

There are 17 possible interactions among these three complexes. BNFL has constantly said a serious accident involving one of their ships is highly improbable. Nevertheless even the Safety and Reliability Directorate — a subsidiary of the UKAEA — recognises that such an accident could occur. It was commissioned by the local authorities to do a hazard analysis report on the gas terminal. The interim report states 'We believe that a substantial spill of condensate during loading could possibly result in fire engulfment of a ship unloading spent nuclear fuel flasks at the BNFL facility within the same basin.' It adds that such risks are 'negligible'. To whom? The people of Barrow were never consulted on this matter, and it must be asked whether such a group as the SRD can really do an independent analysis of safety standards. Also, BNFL does not own any shielded vehicles, which would have to be used to approach a nuclear waste ship on fire ... Should any accident happen to a flask while in Barrow, the people of the town will have to look after themselves. There are no contingency or evacuation plans in the event of a radioactive release. BNFL stated at a public meeting on 14 February that evacuation plans were not its responsibility but that of the civil authorities. The civil authorities have denied any knowledge of these plans. Barrow is on a peninsula and has only one main road out. This is expected to be the evacuation route for 67,000 people, who will presumably all be trying to fend for themselves in a chaotic situation.

We would rather not need to have any plan at all; we are trying to convince all concerned that these shipments must be stopped from entering our country. In view of the Health and Safety Executive's recent report on the leak at Windscale, it is not surprising that we are somewhat dubious about BNFL's so-called 'spotless' safety record.

The cardinal fact is that the nuclear fission process itself is so inherently powerful and potentially dangerous that it demands absolute perfection in the materials, the technology and the people that contain it. The forces of radiation which could be released through a series of individually 'minor' failures means that there is no such thing as a minor nuclear accident — at best we have a lucky escape.

What is worse is that nuclear engineers have incorporated built-in hazards into current designs which are being actively proposed by the present British government. Highly explosive materials such as liquid sodium (in the FBR) and zirconium (in the PWR) are used *because there is no alternative*. In other words, the nuclear industry is operating and proposing reactors with potentially lethal hazards built into their normal working conditions, quite apart from the engineering failures and human errors which figure so largely in the Nugget File. The fast breeder reactor, intended to be the next generation of reactors after the PWRs, contains such a volume of densely-packed fissionable material that even under normal working conditions (an as yet untested dimension) it approaches 'critical mass' — that is, a density that causes an uncontrollable chain reaction leading to a nuclear explosion. Furthermore, the vast heat generated by the FBR is cooled by liquid sodium, one of the most active substances known to science, which explodes on contact with air or water — in which FBR waste is cooled and stored.

The small experimental FBR at Dounreay has already experienced an explosion in the high-activity waste silo, a deep shaft 50 yards from the sea into which much of the most dangerous solid waste is placed.

> Around midnight on the night of 9-10 May 1977, the silo exploded. Sodium, which is used as the heat transfer agent in the FBR, came into contact with water, and blew up, hurling a plug said to weigh five tons from the top of the silo and cracking a 15ft square, 4ft high block of concrete surrounding it. Sodium should not have been dumped in the silo with its shaft so close to the sea. As any chemistry student knows, there is a fierce reaction when sodium and water meet. (*Daily Mail*, 8 September 1980.)

In fact, this accident, and serious losses of plutonium and fuel rods, together with worker health neglect, became known to the public only through the investigations of a BBC television team three years later.

The public has been led to believe that the accident at Three Mile Island has been successfully overcome. In reality it presents a continuing and unsolved danger partly created by the growing bubble of hydrogen gas generated by the zirconium cladding of the fuel rods within the reactor core. A high-energy physicist from Fordham University, Daniel Pisello, wrote in an article published shortly after the Harrisburg accident: 'The dangers of zirconium are well illustrated by the events at Three Mile Island. Mechanical difficulties, the details of which are not of crucial importance here, led to a partial loss of coolant, and a partial meltdown of the reactor core.' (*Ecologist*, May-June 1979). As an emergency measure, reserve cooling water was sprayed on to the dangerously exposed and overheated core. Several days later, it was reported that a huge bubble of flammable hydrogen gas had formed unexpectedly inside the reactor vessel. This bubble not only interfered with efficient cooling of the damaged core, but also presented the frightening possibility of a hydrogen explosion whose likelihood increased hourly as the oxygen concentration in the bubble approached a critical level. Such an explosion would have ruptured the containment vessel, precipitating a meltdown and resulting in large-scale and long-term contamination of the atmosphere and the Susquehanna River Valley. Spokesmen for the utility company and the Nuclear Regulatory Commission claimed ignorance on the subject of the origin of the hydrogen bubble, referring to it as a 'new twist' and 'something that had not been foreseen when the reactor was designed'. The next day the bubble shrank and vanished.

The American media carried the story but gave no explanation, simply indicating that the bubble's disappearance had been more rapid than expected. The claims of ignorance and the pretence of mystery on the part of both the utility company and federal experts with regard to the appearance and disappearance of the hydrogen gas are, quite simply, lies. Explanations for these occurrences are freely available in the literature on nuclear engineering and safety, and centre around the use of zirconium alloy cladding. Experts within the American nuclear establishment are privately admitting that they are certain that the hydrogen was produced by the reaction of tons of zirconium cladding with the steam formed in the reactor vessel during the early stages of the accident.

Weeks after the event, the only public reference to the role of zirconium in the production of the hydrogen bubble was in the British press. (Remember that of all the major nuclear powers, only the UK has no PWR's — yet). The 12 April 1980 issue of *Nature* magazine referred to a recent letter to the *Guardian* by Sir Martin Ryle of the Cavendish Laboratory in Cambridge, who stated that a highly dangerous hydrogen bubble should have been predicted as a matter of 'A-level textbook knowledge'. The following excerpt is taken from a standard textbook on reactor safety and is part of a report dated February 1969:

> The chemical reaction of the cladding with steam, which is supplied by the water remaining in the bottom of the primary vessel after the blowdown has three important effects. First, it furnishes *energy* which can increase the heating rate of the core. Second, *hydrogen*, a reaction product, is released to the containment structure. Third, the reaction also *changes the character of the cladding* (i.e. the metal cladding is converted to an oxide), which can affect its behaviour on quenching.

In 1975, Professor E. A. Gulbransen, a former research scientist at Westinghouse (who make the type of PWR advocated by the present government) published a letter in the *Bulletin of the Atomic Scientists*. In it he commented:

After 25 years of research and development work on the chemical and metallurgical properties of metals and alloys used in nuclear power plants, I have come to the conclusion that the current design and materials cannot give us a safe and well-engineered nuclear power plant ... The use of zirconium alloys as cladding material for the hot uranium oxide pellets is a very hazardous design concept since zirconium is one of our most reactive metals chemically ... There seems to be no way to overcome the inherent material problems associated with zirconium alloys and the current design of the reactor ... No back-up or alternative design is available if the present design proves unreliable.

In the face of such informed warnings, why does the nuclear industry persist in using zirconium in its PWRs? For several reasons: as a metal, it has good structural strength; it resists corrosion at the normal operating temperatures of a nuclear plant; and perhaps most importantly, it is relatively transparent to the neutron 'bullets' which must pass through it to keep the reactor going. Stainless steel, by comparison, would absorb almost a thousand times as many neutrons as zirconium does, making it an unacceptably expensive alternative. It is now becoming clear that the extreme costs of nuclear reactors call their safety into question. This is a point we shall return to later.

The nuclear establishment, not surprisingly, prefers to lay the blame for accidents on the station operators. Gary Zimmermen, supervisor of nuclear licensing for Portland General Electric in Oregon, said of the Three Mile Island accident:

There are certain scenarios that are postulated for certain accidents ... those that are postulated do not envisage a hydrogen bubble forming. I hate to put the operators back at Three Mile Island in a bad light, but if they hadn't turned off the emergency core cooling system, it would have kept the core below the threshold temperature, and the zirc-water reaction would never have started.

However, the Kemeny Commission appointed by President Carter to look into the circumstances of that accident pointed out that one of the vital alert signals was hidden by a safety maintenance tag, and the combination of alert signals could have indicated events of various kinds and yet had to be acted upon immediately. The commission also noted that the job of being a nuclear power station operator demands someone who can cope with hours, months and years of boredom watching the routine signals on huge switchboards, and yet also interpret unusual signals and act decisively and critically within two or three minutes — sometimes even less. Moreover, many of the accidents listed in the US Nugget File or reported to the British Department of Energy since 1976 had nothing to do with the station operators, but stemmed from failures of equipment, techniques or materials, often multiple failures.

In the *Guardian* of 30 October 1979, Shoja Etemad, an experienced nuclear engineer with Framatome, the giant French nuclear consortium, described the collapse of the scientific method on which nuclear safety calculations are made:

The *universally* accepted basic hypothesis is the 'golden rule' conceived by the American safety authorities years ago: that there can only be one failure within the operating system at one time, unless there are others resulting from the first one ... yet at Three Mile Island ... according to the safety authorities there were six independent failures, one following another, but not consequential. The US Nuclear Regulatory Commission believes that two of the six failures at Three Mile Island were due to human errors. One was the switching off of the safety injection system by the operator during the transient (accident). But the NRC was not present at the scene at the time, and the statements by the utility officers were contradictory. Considering that plant after plant is now being re-equipped with a more powerful safety injection system, would it not be more plausible to infer that the safety injection system simply failed to cool down the reactor, and was *not* switched off by mistake? But the NRC did not and cannot agree to that conclusion. If it did, it would have admitted authorising the reactor to operate with a deficient safety injection system. It is easier to blame an operator, for he can be seen to make mistakes. The NRC cannot.

Even accepting the NRC's version, there are other problems. There was a relief valve which failed to close. The French safety authorities responded by saying that instead of relief valves discharging water, they used safety valves which would open under the increasing steam pressure. That is true. But they did not make it clear that after the opening of the safety valves there will be a steam/water mixture passing through. Once that happens it can be expected, as happened before in France and happened again only recently, that the valve will not close again. This possibility was put forward in a report which I submitted two years ago when with Framatome. At Three Mile Island the pressuriser water-level indicator failed to indicate the correct water level. At one stage a few years back on a similar reactor in France our indicated pressuriser water level was so wrong that we decided to eliminate it from our safety protection signal. The fact is that in PWR technology at this time we do not have a proven way of determining the water level anywhere in the system during transients (accidents). I suppose this is part of the secrecy which 'protects' the nuclear system. That is why it was unknown to the operator at Three Mile Island and to anybody else who may, perhaps, have assisted with a solution at the time of that crisis.

Meanwhile in France cracks on the steam generators' tubular plates have just started to get some publicity. They were first noticed on one tubular plate last summer at Framatome and Chalom. But checking through others, which had already been through the sophisticated procedure of Nuclear Quality Assurance, we found that the same cracks were present in a number of others both at Chalom and already in power stations, and *also in part of the primary piping close to the junction with the reactor zone.* The tubular plates about 60cms thick should normally stop any contact between the primary coolant and the secondary coolant. Contact would involve hazards.

A team of engineers, of which I was a member, from the related departments was entrusted with the task of testing the cracks of up to 8mm. × 6mm. × 1mm. to see whether they would grow big enough and fast enough to join existing tube holes and thus allow coolant to leak. Our primary findings concluded that they grew most as a result of the frequent normal operational temperature changes of power station load change. The French authorities did not share our concern and gave the go-ahead to use them ... and ordered two reactors with the cracks to be fuelled and started up ... but the unions refused to load fuel. But suppose the two reactors are started up and a loss-of-coolant accident should follow any first failure, would that be considered a *human* error by the safety authorities? ... even though some of our best scientists work in two-phase flow fields and make superb progress every day, we are still incapable of predicting the flow from the reactor vessel allocated to each loop during a transient when steam is present. Since the reactor vessel is geometrically symmetrical, it should be easy. But it is not: indeed at present it is impossible. This flow allocation is crucial to the amount of steam left in the vessel (whether or not the core has become uncovered by the cooling water) ... Nuclear power, which was our hope for a better kind of life by providing an almost unlimited source of energy — the life resource — is instead becoming a crippled regression in science.

An article in the *Ecologist* (April-May 1980) developed this theme:

Water at the temperature and pressure required for PWR operation is extremely corrosive and once the stainless steel lining covering the inner black steel has been eaten through, the fissures will spread with great rapidity, giving rise to the possibility of pressure vessel failure. An accident of that kind could make Three Mile Island pale into insignificance. Similarly breaches of the heat exchanger plate could lead to loss of coolant and pressure with consequent build-up in steam and a possible explosion. According to Etemad, a fundamental problem with PWRs is for the operator to know the level of water in the reactor during an incident of the kind which struck Three Mile Island.

Moreover, even if the controls are wholly automated, so as to eliminate the possibility of human error compounding the incident, the instrumentation itself may lead to faulty interpretation which then automatically guides the reactor to a worsening of the situation. Also, as reactors age beyond ten years not only does their performance tend to deteriorate, but they are subject to increasing radiation leakage. The Commonwealth Dresden reactor, one of the oldest commercial PWRs in the United States, has become too radioactive for routine maintenance. One way round the problem is to flush out the reactor, and the utility believes that PWRs may have to be flushed out at least twice during their lifetime. Repairs and replacements are also problematic because of radiation, and at the Indian Point plant in 1971, it took 700 workers nearly one year to dismantle and refit a heat exchanger owing to radiation levels of more than 10 rems an hour. A similar effort in a conventional plant would take 25 workers two weeks. All the Turkey Point reactors of the Florida Power and Light and Virginia Electric will have to be extensively repaired or replaced. Finding experienced workers who have not surpassed their yearly permissible dose will become increasingly difficult as more and more nuclear plants are built — the times and therefore costs of repairs will increase; thus jeopardising any supposed competitiveness with coal-fired plants.

Before leaving the subject of the safety record of the pressurised water reactors in general, and the Three Mile Island accident in particular, it is worth quoting both the short summing-up of the Kemeny Report commissioned to investigate that accident, and the comment of Sir John Hill, chairman of both the United Kingdom Atomic Energy Authority and of British Nuclear Fuels Limited. The Kemeny Report: 'While throughout this entire document we emphasise that fundamental changes are necessary to prevent accidents as serious as TMI, we must not assume that an accident of this or greater seriousness cannot happen again, even if the changes we recommend are made.' And Sir John Hill: 'The problem with the nuclear industry is not that we have too many mishaps, but that we have not been able to convince the public that, with the very great care taken, mishaps and minor accidents will not escalate into major accidents which could kill or injure large numbers of people — despite the fact that this has never happened. The accident at Three Mile Island was a serious one, the worst that has happened in a nuclear power station, but because of the barriers incorporated in the design of the plant, no employee or member of the public was injured'.

> The US Safety Information Center at Oak Ridge has recently disclosed that of the 2,000 incidents investigated in 1979, no fewer than 32 might have ended in catastrophic core meltdown. The British nuclear industry refuses to make its safety findings available to the public.

These are the kind of problems we will face if we permit the government to go ahead with its intended large-scale purchase of American PWRs. But the British nuclear industry itself does not have an accident-free record — a look at the list of failures and accidents admitted by the industry, both at its reactor sites (the earlier Magnox reactors and the later Advanced Gas-cooled Reactors) and at the reprocessing plant at Windscale, is instructive. In the *New Statesman* (7 December 1979), Duncan Campell reported:

'Calder Hall and Chapelcross were the first British reactors of the Magnox design. All of these have now been discovered to have substantial cracks in their pipework. If the cracks spread through the structure, and allow the cooling gases to escape, this is likely to require the evacuation of the surrounding population. Fortunately, the cracks have not spread. But, alarmingly, many of them appear to have been built into the reactors during their construction, when the need to test every component with critical exactitude was not appreciated or technically possible. One independent safety specialist who has examined the cracks told the *New Statesman* 'We've just been lucky'.

We nearly weren't so lucky in March this year (1979). After

repairing cracks (up to one metre long) in the welding of a gas bellows (of the primary cooling circuit, close to the core) in the Dungeness A Magnox station in Kent, the CEGB, during a routine biennial inspection, pronounced the reactor safe. Only because the Nuclear Installations Inspectorate took the opposite view were more cracks discovered and a three-month investigation undertaken to make sure that the cracks were not going to spread. Further checks are now being made while the reactor is running. The four-day shutdown cost the CEGB £4 million — highlighting the need for independent safety checks. The NII inspectors are bitter that they earn considerably less than even the CEGB's deputy station managers, with whom their responsibility for safety is comparable.

Dungeness engineers have been quoted as saying (*Nature*, 17 January 1980) that the 16-year old reactors are unlikely ever to be used again. Serious steel corrosion in similar early Magnox reactors meant that the operating temperature had to be derated, thus reducing the electrical output. At a new station in Heysham, the NII inspectors noticed that one section of lighter-coloured concrete had been used; this turned out to be 20 cubic yards of defective concrete which, unnoticed, could have caused total failure of the reactor vessel — known to nuclear safety experts as the 'incredible' accident.

Duncan Campbell lists many horrifying failures in reactor construction: defective stress wires close together, gas input pipes buried unnoticed in corrosive water, vital safety checks on welds being faked by manufacturers to cut their own costs. The *Financial Times* of 23 February 1976 carried the following report: 'At Hinkley Point A reactor a fault in the electrical system caused flooding in the main building and cut off the vital supply of cooling water which keeps the reactor safe. At Hinkley Point B (an AGR), when the reactor was within nine days of producing electricity an ominous rattling started in the reactor. It was deduced that a big valve deep inside the reactor was vibrating violently as it tried to stem the rush of hot gases.' But which of the 308 'gags' in the reactor was it? 'The enormity of the problem only dawned slowly,' admits one engineer. Over a period of months they withdrew more and more of the gags. The damage was frightful. The tough stainless steel work was scoured, bent and battered. In some cases, pounds of steel had been chewed completely away.

Sir John Hill said on BBC Radio 4 in spring 1980 that he *could not afford* to disclose the safety records of the AGR. On the same programme, the NII said that the CEGB were equally reluctant to disclose the report on the PWR, although the inspectorate had requested it some time ago. The veil of secrecy surrounding nuclear matters in this country allows the CEGB to make such statements as, 'the maximum credible accident would cause no hazard of excess radiation to the population at large', in the context of their replacement of steel by concrete in reactor vessels, a move which became necessary because of corrosion problems. The NII, though grossly understaffed and underfunded, has demanded mandatory evacuation plans for all nuclear plants, though as Governor Richard Thornburgh of Pennsylvania pointed out in his testimony to the Kemeny Commission: 'When you talk about evacuating people within a five-mile radius of the site of a nuclear reactor, you must recognise that it will have 10- 20- 100 mile consequences, as we have heard during the course of this event.'

The failures in nuclear safety listed above are failures which fortunately were detected before they contaminated workers or our living environment. Actual accidents have taken the authorities equally by surprise. At Windscale in 1957 a serious fire contaminated an area of 500 square kilometres with iodine-131; it was fought for 24 hours before the Chief Constable of Cumbria was notified, and several days elapsed before farmers were told to pour away their contaminated milk. In fact the fire (caused by an error on the part of a highly-qualified physicist) had been burning for two days before the Windscale authorities discovered it; then their own quenching material, liquid carbon dioxide, only increased the flames. The local fire brigade had to be brought in with water hoses. Afterwards, reactors 1 and 2 were closed and stuffed with concrete.

In 1969 a new reprocessing section was opened at Windscale, with an expected capacity of 400 tonnes, but it had only processed 120 tonnes by the end of 1973, when a blow-back of ruthenium-106 contaminated over 30 workers, some heavily, and the plant was closed for good. In 1976 a leak was discovered at Windscale coming from a waste silo containing radioactive material. More than 20,000 gallons have now escaped from the silo into the ground, and thus into the water-table. The then

Minister for Energy, Tony Benn, claimed that this fact had been concealed from him over two months; as a result he instigated a programme by which all nuclear errors or accidents should be reported immediately to the Department of Energy. However, British Nuclear Fuels Limited and the Health and Safety Executive have still not found a way of plugging this leak. It is thought that the accident was caused when the magnesium alloy containing the fuel reacted with the water in which it was stored, and became so hot that it cracked the storage silo's concrete base. Daniel Pisello has pointed out that the spent fuel rods from PWRs are kept in:

> thin zirconium tubes filled with radioactive waste including plutonium. These rods are stored at the plant site under water in circulating pools designed to carry off the decay heat. A typical pool may contain a ton or more of relatively volatile plutonium oxide. Only a few feet of water separates the flammable zirconium from air in which it may ignite at around 1400°F. A zirconium fire in a spent fuel rod storage pool is one of the worst conceivable disasters, because tons of plutonium would be released into the atmosphere. Every year nuclear reactors in the US produce ten tons of deadly plutonium packaged in a thin cladding of flammable zirconium.

It should be remembered that the PWR so dependent on zirconium is the reactor most favoured by the government for a stepped-up nuclear programme. Meanwhile, at Windscale, the accidents continued.

In July 1979 fire broke out in the fuel de-canning cave; eight workers were contaminated above the permissible levels, and the area was closed for four months. In March of the same year another leak was discovered at Windscale. This had gone undetected in a 'comprehensive' survey of the plant's safety carried out in 1976, and continued ever since. In a report issued on 1 August 1980, the Health and Safety Inspectorate condemned the safety standards and the professional judgement of British Nuclear Fuels Limited over the leak, in 'the strongest attack ever made on a public utility' (*Guardian*, 2 August 1980). Although the report, which BNFL acknowledged to be fair and accurate, said that the management committed breaches of the Nuclear Installations Act, 'only the previous good record of the plant and the prompt remedial action taken by BNFL saved it from a prosecution for negligence' (Norman Lamont, Under-Secretary for Energy).

In addition to the failure to find the leak in 1976, the report, compiled by the Nuclear Installations Inspectorate for the Health and Safety Executive, also points out that the leak could have been stopped in May 1978, when the company failed to follow up the discovery of radioactive contamination outside the building where the leak was eventually discovered. The investigation confirmed that a massive amount of radioactivity — about 100,000 curies — had seeped into the ground beneath a disused building, B.701. Although the 1976 survey failed to spot the leak it 'did notice that the instrument panel in B.701 was unserviceable and should be replaced, and after its report the gauges measuring the level of liquid in the tanks were renovated and recalibrated. But, incredibly, those responsible for the job and for taking regular gauge readings were apparently unaware that the needles had gone right round the dial and were on their second circuit' (*Observer*, 3 August 1980).

And so the list goes on. The newly-opened plant at Hunterston B in Scotland was inexplicably flooded with sea water in 1977, which meant closing the plant down for two and a half years and replacing all the parts corroded by brine. At Dungeness B, already ten years behind schedule in 1979, a leaking joint on a boiler manhole cover allowed radioactive CO_2 to escape.

The Magnox reactor at Trawsfynnydd has been closed since water was discovered in the reactor core, and local grass and therefore milk found to be contaminated. Anthony Tucker, reporting on the nuclear faults revealed by improved techniques of detection (*Guardian*, 9 October 1980) ended:

> The essential safety problem is not simply that the failure of a primary cooling circuit in Magnox reactors could constitute initiation of a maximum credible accident, but that, in spite of public and internal technical pressure for changes in safety procedures, the Hinkley Point reactors are operating without detailed inspection information on eleven of the twelve cooling circuits, and without any technical assessments of the situation by a safety committee set up to regulate problems of this kind.

Accidents in other countries have, in some cases, been more

catastrophic. In 1961, at Idaho Falls in America, there was an explosion from unknown causes so serious that the three workers who died were irradiated to the point where their bodies had to be dismembered, certain parts being buried in lead coffins, the rest being left in the power station, which is to be *guarded for ever*. At the Browns Ferry plant in Virginia, routine maintenance check-ups of the electrical system caused such a serious fire that all electrical safety systems were knocked out and the whole reactor surveillance core threatened. (Similarly-caused fires have occurred at Hinkley Point in Somerset).

At Cap de la Hague in France in 1970, five years after the reprocessing planted started up, radioactive gas in the storage tanks ignited spontaneously and huge bubbles exploded, contaminating the entire building. The workers had to wear diving suits during their maximum working time of three minutes. In 1973 a new plant was opened on the same spot without essential safety equipment; there was an explosion with resultant contamination and workers had to be brought in to deal with the emergency, using the human 'sponge' principle previously referred to. Still at Cap de la Hague, a high-activity waste reprocessing plant was opened in 1976, but had to be shut down to improve radiological protection for the workers, who meanwhile went on strike. When it was reopened in November 1977 it was expected to process 400 tonnes a year, but has barely achieved 120 tonnes. In December 1977 a dissolver became blocked, causing a pile-up of 100kg of fuel, which is very close to causing a critical (explosive) accident. In early 1978 this plant, like its Windscale counterpart, was converted back to reprocessing only Magnox fuel, soon to become obsolete as the Magnox reactors are retired and decommissioned.

In Japan in 1974, the Harbour Workers' Union refused to carry out maintenance work on the nuclear submarine *Mutsu*; in 1977, also in Japan, an oxide reprocessing plant was commissioned to reprocess 200 tonnes a year, but by August 1978 this Tokai Mura plant had broken down after handling only 19 tonnes of waste.

The Nugget File revealed design faults in safety wiring systems in American reactors: floating switches to ensure emergency shut-down were found to sink, and in one plant there was a danger alert when someone 'accidentally bumped' a theoretically earthquake-proof switch. In California an inefficient reactor had to be sliced up and buried in the Nevada desert, at a cost of at least $6 million. In Switzerland, the underground reactor at Lucens suffered a loss-of-coolant accident, blew up, and had to be scrapped and covered in. In Sweden, the Marriken reactor simply did not work and was converted into a conventional power station.

In Detroit in October 1966, the Fermi fast breeder reactor put out a civil defence and SCRAM (close-down) alert as the core fuel melted. Walter Patterson has commented: 'The FBR has its fuel in a configuration which may be considerably short of ... its maximum theoretical activity; if the Fermi core had been distorted and melted it might have been susceptible to local surges of radioactivity, intense hot spots which could lead to chemical reactions between fuel, cladding and coolant, and even to violent chemical explosions which could cause a fully-fledged nuclear explosion.' The dangers of the fast breeder reactor will be discussed more fully in a later chapter, but this Fermi accident is significant to us in the context of the British government's proposal for a fast breeder programme for this country.

We return finally to the accident at Three Mile Island, important because of the government's interest in pressurised water reactors, but also because it took place in a reactor that was not in fact working at full capacity. Whatever the reasons for that accident, the problems it has caused and the dangers that it represents will clearly remain inescapable for many years. The Kemeny Commission found that the danger point of such an accident arrived much sooner than had been calculated; the heat tolerance accepted by the industry had been set far too high. The building cannot be decontaminated until 1983 — if then. Metropolitan Edison, the operating company, are unable to carry out normal repairs and maintenance on number 1 reactor building (where the emergency core cooling system failed), and therefore cannot replace any of the seals or glands which are deteriorating. The level of radioactive contamination still present in the vast amount of water in the building would be fatal to anyone entering. Added to this, seasonal changes in temperature alter the pressures inside the containment building and impose greater strains than normal on the seals, which by now would have been routinely changed. They may give way and release radiation in the form of krypton-85, a gas with a radioactive half-life of ten years, which is present in large quantities. Ninety tons of uranium fell to the bottom of the core housing, the whole interior being severely damaged in a temperature of 3,900°F. The krypton gas could be deep-frozen and removed, but that would add at least 20 months to the decontamination timetable. The alternative would be to use charcoal to absorb the radiation, but that would call for one-third of all the charcoal available in the US.

As a result, Metropolitan Edison have asked for permission to release the gas; workers will be subjected to additional doses (several are already in excess), and the Kemeny Commission considers that there is still some risk to the general public. 'Ahead lies a decontamination effort unprecedented in the history of the nation's nuclear power industry — a cleaning-up whose total cost is estimated at 80 to 200 million dollars and which will take several years to complete' (the Kemeny Report). This report also noted that the release potential still existing inside the reactor would demand large quantities of medical-grade potassium iodide to dose the general public, who are already at risk both from the radioactive cloud emitted from the vents, and from the radioactive water which overflowed into the Susquehanna River from the tanks. Unfortunately, no chemical or pharmaceutical company is at present marketing potassium iodide in the quantities needed. Once again, it is clear that the resultant costs of this accident, which was not 'the worst possible', were neither anticipated nor prepared for.

> 'We were diverse in our expertise, in our backgrounds, in our traditions, and in our beliefs on nuclear power. Our unanimous verdict may well have been the miracle of 1979.' Professor Kemeny, on the Kemeny Report on Three Mile Island.

A final, so far apparently unconsidered, hazard potential in the nuclear power process is the vulnerability of the cooling ponds (which contain high-level radioactive waste outside the reactor containment) to quite ordinary low-level explosive bombs — producing the effect of nuclear bombs. In the early stages of the Gulf War between Iran and Iraq, an Iraqui nuclear power plant was bombed by either Iran or Israel. The consequences of this attack have not been fully made public; the possible results are horrifying to contemplate.

4. The FBR and the Plutonium Cycle

At the beginning of the nuclear power programme, uranium was relatively cheap and thought to be in more or less unlimited supply. At present, it is calculated that there are approximately 2.5 million tonnes of available uranium throughout the world. However, so many countries have now instigated civil and military nuclear programmes that, by the year 2000, demand will exceed known reserves.

During the summer of 1980 it became known that Rio Tinto Zinc and various Canadian transnationals intended to mine for uranium in Donegal, in the Republic of Ireland. Donegal is an extremely rural county and opposition has been vociferous, particularly after it became known that, under the EEC Treaty of Accession, any uranium found in Donegal would belong to the EEC. The opposition to mining in Ireland shows that the local population has become aware of the deadly effects of the radon gas released during mining, and from the uranium tailings; it has now been established that there is a 40 per cent increased chance of lung cancer among uranium miners and those living near the mines.

In Canada a full-scale political controversy has arisen over the large uranium deposits in Saskatchewan, which many people would prefer to leave unexploited; meanwhile the federal government has imposed export restrictions on uranium. The Australian government has likewise imposed restrictions.

An acrimonious debate is developing over the British purchase of uranium from Namibia. Rio Tinto Zinc signed the original contract with South Africa for Namibian uranium on behalf of the UKAEA in 1970 — and failed to inform the Cabinet that it had done so. Although the Labour government agonised over the decision, it did not in fact revoke the contract. Not surprisingly, the Namibians regard the contract as international theft, and guerrilla war has broken out around the mines and the export routes. Ten years later, in the context of a UN decree declaring South Africa's occupation of Namibia illegal, and with the governments of the world moving towards sympathy with the claims of Namibia and SWAPO, the present government has refused to review the contract (*Guardian* 21 August 1980).

Such political problems, together with the scarcity of uranium, mean that existing reactor designs will have a very limited life, particularly in those parts of the world which do not have an indigenous uranium supply. Any medium- to long-term nuclear power plans must replace uranium as primary fuel. The fast breeder reactor seems to be the perfect technological answer, in that its fuel is plutonium, the waste product of the uranium reactors; and in that it is theoretically capable of actually producing more plutonium than it uses, thus allowing the cycle to expand.

In *World Watch* paper no. 6, Denis Hayes wrote:

> Without breeder reactors, known uranium reserves will not long support nuclear development ... As prices rise, low-grade recoverable ore can probably be found, but at high prices, not competitive with coal, and about equal with oil. (It also takes approximately ten years to produce fuel uranium from a deposit, once it has been located.) Only with breeder reactors can we derive sufficient energy from otherwise unusable U-238 and thorium to make such low-grade sources attractive. Current-model breeders, however, are designed to maximise the production of plutonium.

Terence Price, head of reactor development at Winfrith, has written:

> Without breeding, at least to European eyes, nuclear power would be merely a transient phenomenon (*Atom*, no. 265, November 1978).

The countries with large nuclear programmes which possess sufficient deposits of uranium are the US. Canada and France; those which do not are the UK, Germany and Japan. The countries with nuclear weapons are the US, USSR, UK, France, India and China; it now appears likely that Israel and South Africa can also be included in this list which, thanks to the developed world's export of civil nuclear technology, is growing yearly. To produce the plutonium necessary for nuclear weapons these countries need reactors and reprocessing plants — that is why reactors were first developed. Brazil and Pakistan are arranging to buy reprocessing plants from France and Germany; France and the UK reprocess separated plutonium from other countries. As uranium becomes scarcer, the attention of all these countries will inevitably turn towards the fast breeder reactor, with its theoretical ability to generate plutonium. Yet as George Wyle, who was Enrico Fermi's assistant in December 1979 said, the decision to proceed with an FBR programme will absorb millions of pounds, with no return on capital in the short or even medium term. The full benefits of the FBR (if they materialise, which recent French experience seems to put in doubt) will not be realised until about the year 2040, although work first started in earnest on the FBR in 1950. He added that the huge sums of development money would be better spent on benign energy sources, (to which he has now dedicated himself) which could take less than 20 years to develop, and on perfecting alternatives.

The doubtful operation of fast breeder reactors, and their so-called 'gains' of plutonium, are described by Peter Bunyard in the *Ecologist*, April-May 1980:

If Super-Phenix (the French commercial 1,240-megawatt FBR newly introduced at Creys Malville) works according to predictions (based on experience with the Phenix, a 250-megawatt prototype like that at Dounreay), it will breed some 165 kilograms of plutonium each year out of an initial charge of 5.5 tonnes — the plutonium gain being 3 per cent ... According to Walter Marshall, deputy chairman of the UKAEA, the fast breeder is something of a misnomer in that its production of plutonium per gigawatt-year is less than other reactor systems, including PWRs. Thus a Magnox reactor will yield 600 kilogrammes of plutonium per gigawatt-year, and a PWR 275 kilogrammes. In fact there is an actual overall consumption or incineration of plutonium in the core of a fast reactor, and that loss is only made up by the breeding which goes on in the radial and axial blanket regions. According to Marshall, out of 2.8 tonnes of plutonium that are used to fuel a 1-gigawatt reactor, some 220 kilogrammes are consumed in the core, while 409 kilogrammes are gained in the blanket. Maximum production with the blanket in in place is thus 189 kilogrammes per gigawatt-year, a figure higher by 13 per cent than the anticipated figure for Super-Phenix with its rather larger plutonium inventory. But to talk of plutonium production in a fast reactor and actually to make use of bred plutonium is another matter. Reprocessing of FBR fuel is a vital, integral part of the fast reactor concept, not only to extract the plutonium bred from the blanket but also the considerable quantities, accounting to several tons, still unconsumed in the core ...

Experience with reprocessing fast reactor fuel is very limited, approximately one ton having been re-treated over a period of ten years in France. Meanwhile in Britain a small pilot reprocessing plant has recently been commissioned at Dounreay for treating spent fuel from the prototype fast reactor. The problem with reprocessing is that the technical difficulties and the risks of serious accidents increase commensurately with the higher burn-up of the fuel ...

Fast reactor fuel, with a burn-up of 80,000 to 100,000 megawatt-days per tonne, is many times more radioactive than thermal oxide fuel (AGRs and PWRs); moreover it has a high content of plutonium, with added dangers of criticality (explosiveness). Any reprocessing of FBR fuel is still very much in the research and prototype stage, with the Commissariat à l'Energie d'Atome in France trying out a dry process in addition to the conventional Purex method ... in this wet process insoluble particles, likely to capture plutonium and possibly bring about a criticality accident, are a particular problem. The particles have to be captured on filters leading to an activity of up to 10,000 curies per kilogramme with a heat discharge of up to 10 kilowatts. Aside from a special geometry to prevent accumulation of excess plutonium (leading to criticality), special neutron absorbers are used in the solvent, the best available being rare earths such as hafnium and gadolinium, which are by definition scarce and expensive. In time these become poisoned by neutron bombardment and have to be replaced.

The handling of fast reactor fuel once out of the reactor is technically as demanding as the safe operation of the reactor itself. At the moment of discharge the residual activity in each fuel element amounts to 30 kilowatts, and for a period of up to 20 days, while the spent fuel is contained next to the reactor, this heat must be dissipated. Any breakdown in the cooling could cause the fuel elements to melt, giving rise to an accident at least as bad as that of a nuclear excursion in the reactor itself. At the end of that period the residual activity has fallen to 7.5 kilowatts on average, and the fuel assemblies are taken out and immersed in either liquid sodium or a sodium potassium alloy in special metallic containers. As with the sodium coolant used in the fast reactors, all contact with air and water must be avoided (any such contact being highly explosive).

The operators of Super-Phenix reckon that some 20 tons of sodium coolant will be required each year for the transport of the fuel assemblies from the reactor to the reprocessing plant, the containers themselves being contained in 60-ton fortified casks. The sodium will meanwhile become contaminated (and heated) by leaking fuel elements and by neutron activation. Again any loss of coolant could be disastrous. In distinction from thermal reactors in which the fuel must be in its most reactive configuration, the fuel in fast reactors is not in its most reactive configuration ... and can therefore melt into a more reactive configuration, thus giving rise to a nuclear excursion (accident). Meltdown of the fuel elements of the FBR, whether inside the reactor, during transport, or while waiting for reprocessing, carries considerable dangers.

The fate of nuclear power hangs on the efficiency, safety and economics of reprocessing. The rationale for the FBR is that it will both increase the total energy extracted from uranium by a factor of 60 or more above a purely thermal programme, and that any country which has accumulated a large stockpile of uranium can, by dint of plutonium, become independent of uranium suppliers. For that reason Giscard d'Estaing talked of France attaining 20 fast reactors by the year 2000. But the efficiency of reprocessing and its safety are very much in question, an indication being the substantial discharges of radioactive waste, including plutonium ... At Windscale, plutonium losses are estimated at more than 7.5 per cent ... and may after fuel fabrication be as high overall as 12 per cent. Assessments of the plutonium losses incurred during reprocessing and fabrication of high burn-up thermal oxide and FBR fuel indicate that they may be equally high.

The breeding of plutonium in fast reactors is usually assessed in terms of doubling time. A doubling time of 30 years thus implies that, operating for that period, a fast reactor would have generated sufficient plutonium to fuel both itself (or its replacement) as well as another reactor. The doubling time depends first and foremost on the amount of plutonium bred in the reactor — the plutonium gain — but also on the rapidity with which the plutonium can be got back into the reactor after reprocessing and fuel fabrication. If Super-Phenix produces a surplus of 165 kilogrammes each year out of 5.5 tonnes of its initial plutonium load, then the doubling time would be just over 30 years. If however it takes one year to reprocess and fabricate spent fuel then the doubling time of a single reactor increases by 50 per cent, if two years by 100 per cent and so on, that is assuming the fuel stays in the reactor core for two years.

With a programme of fast reactors and the implementation of a programme of reprocessing to keep pace with spent fuel, then the overall doubling time of the reactor group will be 0.7 of the doubling time of the single reactor. What about plutonium losses? If these amount to as much as 6.5 per cent then they would cancel out completely any plutonium gains and extend the doubling time to infinity, at least in Super-Phenix in which 165kg of plutonium are bred in a total of 2.5 tonnes of plutonium taken out of the reactor at the end of a year's operation. Any losses above 6.5 per cent would mean that fuel for the working FBR would have to be made up from an outside source — a thermal reactor (AGR or PWR).

To keep the total inventory of plutonium as low as possible in a fast reactor programme, the operators must try and reprocess spent FBR fuel as soon as they can after extraction from the reactor core. Indeed with Super-Phenix 2.5 tonnes of plutonium are bound up in each refuelling operation, an amount obtainable only after 10 years of PWR operation. The residual radioactivity in the spent fuel diminishes rapidly after extraction from the reactor, although after ⅓ of a year cooling, its gamma activity is still 50 times higher than that of the spent PWR fuel which has been cooled for five years. The hope is to be able to recycle the plutonium in less than a year, including fuel cooling time, reprocessing and fabrication.

To achieve this rapid recycling, Chauncey Starr of the US and Walter Marshall of the UKAEA have proposed a completely automated reprocessing system which allows certain fission products to pass through with the stream of uranium and plutonium. The reprocessed fuel could therefore contain fission products with a gamma-activity level comparable to that of five-year-old PWR spent fuel. Fuel fabrication would also have to be automated and Marshall has proposed a gel process now under investigation at Harwell. Whether such a process — termed the Civex process — would work efficiently and safely at an industrial level is simply not known, nor has it even been tried out at the laboratory level. The future of the FBR would therefore seem to be on very shaky ground. The methods of reprocessing tried at the industrial level give rise to large plutonium losses which may exceed the gains made in the reactor. And in order to recycle the fuel promptly after extraction from the reactor, fuel will have to be handled with very high radiation levels. Proposals for automating fast fuel reprocessing are at the early experimental stage, and may prove impossible to scale up. What would happen in an automated plant if radioactive residues build up to dangerous level? Would that situation have to be overcome by building in multiple redundancy, and at what cost?

Peter Bunyard deals at length with the reprocessing of FBR fuel because he considers that reprocessing plants are possibly even more vulnerable to major explosions than the fast breeder reactors themselves. As he points out, the plants have not been designed and built to withstand large explosions and massive radioactive discharges. 'At the Cologne Institute of Nuclear Safety in 1976, some engineers carried out some studies on the gravest accident they could envisage for a reprocessing plant, and they concluded that a radiation cloud, thousands of times bigger than Hiroshima, could envelope an area downwind from the plant and kill many millions of people.' He points out the dangers inherent in the operation of the fast breeder reactors themselves as follows:

No full size fast reactor has yet been in operation. Whether such reactors can be made acceptably safe is a point of major controversy. A number of eminent physicists at CERN, for example, believe that the technology should be abandoned on account of safety alone. Besides five tonnes of plutonium in the core, the fast reactor contains 6,000 tonnes of liquid sodium coolant at a temperature of between 400°C and 600°C. In contact with air such sodium burns spontaneously; in contact with water it produces hydrogen and caustic soda. Sodium fires have occurred as at the Shevtchenco fast reactor in Russia, and safety problems in general have made the Russians cautious about the rate of expansion of their fast reactor programme. As it happens, no technology exists at present for controlling sodium fires which involve more than a few hundred kilogrammes.*

Should sodium leaks occur in the secondary heat exchanger, causing hydrogen formation, then automatic safety devices come into operation to shut off the coolant circuit and to flare off the hydrogen, like burning gas from an oil well.

Although sodium is not corrosive in the same way as pressurised water, it has a high affinity for carbon which it drags out of the steel reactor vessel and fuel cladding, bringing about embrittlement and a loss of structural properties. Ageing effects in fast reactors must therefore be watched very carefully. The Doppler effect, whereby a rise in temperature in the reactor brings about a fall in reactivity, is extremely important as a means of control in fast reactors. Nevertheless situations can be envisaged in which their chain reaction becomes runaway and uncontrollable. One possibility is that the sodium pumps fail. With the reactor shut off, natural circulation of the sodium by convection should take care of residual heat in the reactor, so far so good but what would happen if the

---

* cf. the explosion at Dounreay, p. 40.

control rods fail to fall into the core — a probability evaluated at one chance in 10 million a year? Within ten minutes the sodium begins to boil followed, at a temperature of around 1700°C, by the cladding around the fuel melting. The interaction of relatively colder sodium and hot molten fuel could provoke a great release of energy — estimated by the CEA at 560 million joules. The vaulting and containment dome above it are supposed to withstand an explosion of 800 million joules, or the equivalent of 200 kilogrammes of TNT. According to the Centre National de la Recherche Scientifique (Commission 06): 'certain critical elements, such as the reactor vessels of Framatome or the vaulting of Super-Phenix, have not been designed on the basis of safety coefficients but purely on the basis of what is technically feasible.' The ability of the vaulting and dome to take the maximum credible explosion is therefore suspect.

Other possibilities exist whereby a dangerous nuclear excursion can arise in fast reactors. Since the core is not in its most critical geometry, a shifting of nuclear material in the core could give rise to an excursion. Such changes could be brought about by a misplacing of a fuel assembly, by a rupture of fuel cladding, by the ejection from the core of a control rod through sodium vapour pressure, or by the deformation of the core hindering the movement of control rods. In fact reactivity can increase extremely rapidly in a fast reactor — at a rate closer to that of the chain reaction sustained in an atomic device than that found in thermal reactors. Should the coefficient of multiplication pass a certain threshold the power of the reactor will double very rapidly, multiplying by as much as 1,024 in a millisecond. A number of explosions could result, with the breaching of all containment. At the end of a year's operation a fast reactor of the size of Super-Phenix contains between 10 and 25 billion curies.

5. Costs

Before examining the economic costs of nuclear power, we should remember the environmental price paid. Nuclear power stations have to be stituated away from densely populated areas, and also need somewhere — like the sea — to dispose of their low-level waste. In effect this has meant that they have been sited in some of our most beautiful and hitherto unspoiled country areas; the Lake District (Windscale), the Welsh coast (Wylfa in Anglesey), the North Somerset coast (Hinkley Point), the Kent marshes (Dungeness), the Suffolk countryside (Sizewell), the Scottish coast (Hunterston), and even John o'Groats (Dounreay prototype FBR). The new proposals for building PWR reactors in the south-west of England have earmarked three sites on the Tamar, one on the Cornish coast at the National Trust beach of Gwithian, further installations at Hinkley Point, and a new installation on the most beautiful Dorset coastal bird sanctuary of the Chesil Beach. The barbarian demands of nuclear power stations are: a clear level area of 200 acres with good foundations close to an abundant supply of water; easy access to road, rail or sea transport; short connections to existing power transmission lines. It was presumably in order to service a future nuclear power station (for which permission had not then been sought but is now being requested) that Dorset was completely bisected in 1965 by the enormous, dangerous and hideous lines of the National Grid, despite strong protests from Dorset County Council.

One would think that the decaying docklands and wastelands of our inner cities would provide ideal sites for nuclear power stations, if they are as safe and as clean as the authorities would have us believe. The CEGB explains its choice of sites in terms of population densities and the consequent complexities of evacuation plans. Apart from revealing the ambiguity of the official line on safety, this insistence on remote sites has two economic consequences.

Firstly, as the power is generated well away from the centres of consumption, there is no possibility of using the waste heat for district or low-temperature factory heating. All power stations, whether coal, oil or nuclear, are about 30 per cent efficient; the remaining 70 per cent escapes up the cooling towers or into the cooling water. Urban power stations could increase their efficiency by operating as combined heat and power stations, common on the continent, although, inexplicably, in this country there is little official encouragement for such schemes. Secondly, the farther the power station is from its eventual consumers, the greater the power lost in transmission. In 1976 transmission and distribution losses amounted to 7.9 per cent of electricity generated; a large programme of remote nuclear power stations would significantly increase that proportion.

However, perhaps the most ironic aspect of the environmental toll exacted by nuclear power is the fact that nuclear power stations *will not go away*. We do not know how to decommission — i.e. remove — them. The originally projected lifespan of the Magnox reactors was 20 years (although two of them have been given extensions despite cracks and corrosion). In fact, the land requisitioned for them will remain unusable for up to 500 years. In an article in the *Sunday Times* of 2 November 1980, Roger Ratcliffe reported:

> Up to two dozen huge concrete cubes, towering 100 ft high and standing in fields ringed with tall fencing, will appear round Britain's coast by the year 2000. They will form gigantic tombs encasing the highly radioactive cores of retired atomic power stations which the nuclear authorities now realise are too dangerous to demolish ... the chairman of the West Cumbrian local liaison committee's environmental group says 'there are already some spots in the Windscale and Calder Hall works which can be described as "frozen". No one will ever walk there again.' Maurice Telford, local Friends of the Earth spokesman, expected an announcement about the closure of Calder Hall a couple of years ago. 'I keep hearing that it is to close, but the announcement never comes,' he says. 'When I first went round the works over twenty years ago I was really impressed by the idea that here at last we had clean electricity. With areas contaminated by leakages, and reactor sites now obviously out of bounds for generations, I have changed my mind. What we've got is a nuclear graveyard here'.

In addition to the environmental drawbacks of nuclear power stations, the straightforward cost has been so astronomical that the CEGB and BNFL have never presented coherent accounts to the public. The whole programme has demanded enormous capital investment in research, construction, operating procedures and safety controls (or cleaning-up after failure). The UKAEA table of expenditure, 1978-79, confirms government subsidies on a scale which coal or oil-fired power generation could never hope for; nuclear costs can never be considered competitive when propped up by accounts like these (see p. 64) It is interesting to note that the FBR alone claims £70 million of these subsidised costs; in 1980 the government increased the UKAEA publicity budget from £350,000 to £650,000. It is not hard to see why.

UKAEA expenditure 1978-79

	£(millions)
gas-cooled systems	7.5
water-cooled systems	6.5
Dounreay prototype fast-breeder reactor and plant	28.4
nuclear materials for FBR operation	17.8
FBR development	23.8
safety	17.4
radioactive waste management	7.4
remaining power programme work	3.4
other costs	1.2
research into fusion	13.9
underlying research	12.3
total	139.6

Costs rechargeable to customers

reactor services	11.6
nuclear research and development and services	13.9
electricity sales	8.6
non-nuclear research and development	14.7
other goods and services	17.3
total	66.1
deficit	73.5

The CEGB's accounts, due to their unusual (and unexplained) accounting and costing methods, remain opaque to the general public, who nevertheless foot the bill. In 1979 Sir Francis Toombs, then chairman of the Electricity Council, told BBC 'Panorama' (when being pressed to enlarge upon the true costs of nuclear power stations) that there are some things the public just has to take on trust, and we would have to take his word for it that nuclear juice is cheap (Jeremy Bugler, *Vole*, November 1980). However, a highly instructive table of costs and over-runs in the new thermal AGR programme intended to replace the obsolescent Magnox programme has recently come to light. This table was sent to Arthur Lewis. MP. by the CEGB chairman, Glyn England. in reply to Parliamentary Question 484 (Q108W) asked on 27 July 1979 (see p. 66, 67). It should be pointed out that most of the labour-related delays mentioned stem directly from the design 'changes' also referred to. For example, Dungeness B was scaled up 20 times from its Windscale prototype, a procedure which all engineers know to be both unpredictable and potentially dangerous. Three different designs were incorporated into the structure as building proceeded, and, in order to save money, the construction was done on site. One result of the scaled-up misproportions and design differences was that the space between the inner and outer containment vessels, in which vital welds had to be made, was later discovered to be too small for an average man to get into. Very small welders had to be recruited — but this begs the question of whether there are any inspectors from the Nuclear Installations Inspectorate small enough to get inside and check the safety of welds?

> 'The AGR programme is a disaster we must not repeat.'
> Sir Arthur Hawkins, ex-chairman of the CEGB.

No wonder the CEGB, although an uncontested monopoly, had a publicity budget of £2.5 million for 1980, excluding its regional activities; in 1979 the Electricity Council spent £4.8 million on publicity and exhibitions. However, despite these efforts, other costs still remain unclear. For one thing, we can only speculate

Estimated final cost of AGR stations
(excluding nuclear fuel)
£ million

Year estimated costs at March price levels	Dungeness B	Hinkley Point B	Hartlepool	Heysham
1965	89			
1966	94	95		
1967	97	94		
1968	99	97	92	
1969	101	101	94	
1970	103	106	112	
1971		113	118	142
1972		116	148	168
1973		124	153	173
1974	200	133	174	191
1975	249	138	200	226
1976	249	143	220	242
1977	319	153	281	297
1978	377	160	365	381
1979	410	160	381	396

Schedule 2: August 1979
Note: Escalation of costs has occurred because of design changes and increases in time-related costs resulting from industrial disputes and other labour difficulties further worsened by inflation.

AGR power stations:
First supply of electricity to national grid

Power Station/Unit	Year of start on site	Contractors'* original programme date for first supply of electricity to grid	Actual or current estimate of date for first supply of electricity to grid
Hinkley Point B	December 1967		
First reactor (R4)		April 1972	February 1976
Second reactor (R3)		October 1972	October 1976
Dungeness B	January 1966		
First reactor (R21)		May 1970	mid-1980 (est)
Second reactor (R22)		May 1971	mid-1981 (est)
Hartlepool	December 1968		
First reactor (R1)		September 1973	early 1981 (est)
Second reactor (R2)		June 1974	late 1981 (est)
Heysham	December 1970		
First reactor (R1)		September 1975	early 1981 (est)
Second reactor (R2)		April 1976	late 1981 (est)

Schedule 1, August 1979
* Turnkey contracts (ie a single contract for supply of the whole of the completed power station) were placed with: TNPE (The Nuclear Power Group), Hinkley Point B; APC (Atomic Power Construction, a wholly owned subsidiary of CEGB since 1976), Dungeness B; BNDC (British Nuclear Design and Construction), Hartlepool and Heysham.

about what the *interest* on the vast and appallingly unproductive AGR investment might be; it is not specified, possibly not included, in the above tables. We do know, from the answer to a parliamentary question recorded in *Hansard*, 2 July 1979, that the cost to the CEGB of making good the electricity lost due to the safety shut-down of Dungeness A amounts to £1.5 million a week. We also know that the estimated cost of the proposed thermal reprocessing plant (a 1978 estimate) was £17 million, but that some of this amount has been guaranteed in advance by the Japanese government, which has no desire to fill Japan's land and oceans with radioactive waste (the British public, however, was not consulted on this bargain so damaging to their environment).

We know that a uranium enrichment plant in Ohio consumes 10 per cent of the state's electricity (coal-fired), more than the whole city of Cleveland. (*Ecologist*, October 1979). We have not been told whether the CEGB are including such costs in their projections for the future when the ageing Magnox reactors are phased out, but we do know that the present costing of nuclear electricity, comparing it to oil- and coal-generated electricity, is based on Magnox stations only.

The following figures were given by Glyn England in a letter to the *Observer*, 4 November 1979:

Cost	(p/kWh)
Nuclear (Magnox)	1.02
Coal-fired	1.29
Oil-fired	1.31

Mr England said: 'The figures relate to stations commissioned since 1965, and include only the 13 most recent coal-fired stations, two oil-fired stations and six Magnox reactors. The costs for nuclear stations include provision for reprocessing, and vitrification of residues from nuclear fuels, and for the ultimate decommissioning of the stations.' It is interesting that vitrification can be costed before it has been perfected, and that the decommissioning of a reactor can be costed at a time when no reactor has yet been decommissioned in the UK.

Roger Ratcliffe wrote in the *Sunday Times* 2 November

1980:

> The CEGB recognised in the mid-seventies that it had nothing put aside for the demolition and has now accumulated a special fund totalling £65 million. But even this, with future contributions of around £15 million a year, is unlikely to pay for much of the work. In the United States, where some successful scrapping of much smaller reactors has occurred, the cost of the operation has been as much as the original cost of construction.

Could this be one of the reasons for the sudden rise in electricity charges during 1980? And as for vitrification costs, do these also include the cost of drilling preliminary holes for dumping high-level waste, the cost of excavating enormous undergound caverns, the cost of permanent security guard over such deadly material — and the cost of persuading the public that such a process is safe and acceptable?

Further, the *CEGB Statistical Yearbook 1979-80* contains some comparative figures which shed an interesting light on the 1979 figures supplied by Glyn England:

Average thermal efficiency
(3,600 as a percentage of heat consumed per kilowatt-hour supplied. 3,600 kilojoules = 1 kWh of electricity)

Coal, gas & oil-fired steam power stations	31.68 per cent
Nuclear power stations	26.78 per cent

Works cost of electricity supplied

	pence per kilowatt-hour (p/kWh)
Coal, gas & oil-fired steam power stations	1.6294
Nuclear power stations	1.1876

Remarkably, the nuclear power station table includes an item not listed in the conventional table — Other Fuel Cost. This presumably refers to the enrichment of uranium — of which costly business, more later. But by far the most striking feature of the nuclear power table is the footnote which instructs us that this so-called 'total works cost' *excludes* 'Insurance, capital charges and provision for decommissioning costs'. It is therefore no comparison at all; and sheds a clear light on the future AGR and PWR costs.

The table giving comparative generation costs is perhaps the most baffling of all, as it claims to include capital charges, interest during construction, provision for decommissioning costs, and *inclusive* fuel costs, and yet still produces the following figures:

	pence per kilowatt-hour (p/kWh)
Nuclear stations	
Magnox	1.30
AGR	1.35
Coal-fired stations	1.56
Oil-fired stations	1.93

This can only mean that it is based on *estimates* of decommissioning costs, and also on current prices for uranium-308 — about US$40 per pound. None of this uranium is available within the UK, the world supply is unknown but certainly limited and, in view of world politics, not to be guaranteed. Are these tables therefore paving the way for forcing the unproved FBR on the public? Or is the CEGB accountant descended from the Red Queen in *Alice Through The Looking-Glass?*

The point about Mr England's 1979 use of Magnox reactors for cost comparisons is that the expensive process of uranium enrichment is not required for these relatively low-temperature reactors: it *is* for the AGRs (if they ever work) and the PWRs currently proposed by the government. Furthermore, cost inflation since the Magnox reactors were built cannot have been covered by these costing methods.

Counter-Information Services, in their Anti-Report no. 22, *The Nuclear Disaster*, comment:

The US Atomic Energy Commission reported in 1974 that the capital costs of nuclear power had risen 500 per cent in the previous five years. In Britain, prices have risen considerably since then: £100 of capital from 1967 is worth (at the very least) £250 now. A realistic costing of power produced by such old plants has to take account of inflation. It is not enough to share out the original costs of the station — the cost of replacing worn-out plant and machinery at today's prices has to be considered too. The CEGB recognises the importance of accounting for inflation in its own accounts; it included a supplement of 40 per cent of its depreciation allowance in 1977-78 to allow for inflation. But in proclaiming the cheapness of electricity from the Magnox stations, the CEGB makes no allowance for inflation ... its published comparisons are based on historic, or original, capital costs ... This means that the capital costs on all stations are underestimated, and because these are such a crucial part of nuclear costs, comparisons are automatically distorted in favour of the Magnox plants. It is made worse by the CEGB's accounting policies which minimize the capital costs in the first place. The Magnox stations took from six to ten years to build. During that time interest had to be paid on the capital costs, before any revenue was earned, amounting to many millions of pounds. This money is itself a capital cost — it is part of the cost of building the station. It ought to be treated as such and shared out over the life of the plant. The CEGB does not include interest during construction as a capital cost.

An independent research group, PERG (Political Ecology Research Group) applied to the government for a grant of £30,000 to undertake a study of FBR safety issues, to present at inquiries into FBR safety. This was refused. However, the West German state of Lower Saxony commissioned PERG to review FBR reprocessing plans, to be presented at a debate with international experts from the nuclear industry. In 1977, speaking at a Royal Institution Forum on 'Nuclear Power and our Energy Future', Michael Posner of the *Economist* said:

> If it is appropriate to have at least two full years' working experience of, say, a fast breeder reactor, before we as a nation commit ourselves to buy dozens of these dangerous animals, then it is a matter of simple arithmetic to show that it is inconceivable that the breeder will make any major contribution to our needs before the second quarter of the next century. Now ... if *everybody* knew this conventionalised fact, the price today would fully take account of these long lead-times: the timelags, the eventual patterns of supply and demand, and the rate of discount would all be connected together.

Dervla Murphy, writing in *Blackwood's Magazine*, recently pointed out: 'A civil nuclear industry could never have been founded on an ordinary commercial basis, because the capital involved is off the commercial scale, the lead-times are too long, and the technology is too complicated.' Or as Walter Patterson (Friends of the Earth) put it rather more trenchantly in a letter to the *Guardian* on 11 November 1979:

> So the CBI is virtually unanimous in its 'backing' for nuclear energy (*Guardian*, 6 November 1980). Does this 'backing' extend to actual money? Would CBI members now perchance pick up the annual tab for the UK Atomic Energy Authority — £129 million last year — or the billion pound over-runs on the first AGR programme, or the ballooning costs of ordering superfluous nuclear stations like Torness and Heysham B to keep the reactor builders from collapse, or the £40 millions for the accident at Hunterston B? Or must the rest of us continue to support the nuclear lame duck in the style to which it is accustomed, while it flounders from one cock-up to another?

Charles Komanoff, testifying in the US Senate, said recently that the rapid rises in the capital costs of nuclear power stations would make it *cheaper* to write off any nuclear power projects which were less than 40 per cent complete and build an entirely new coal-fired plant. The California Energy Commission found that it was probably going to be cheaper to charge for a fully completed nuclear power plant *and also* for use-efficiency and insulation improvements that would render the plant unnecessary, than to keep fuelling the plant.

But perhaps the last word on these astronomical costs should, fittingly enough, come from the Astronomer Royal, Sir Martin Ryle. In a letter to the *Observer* 2 March 1980, he said:

> The government has proposed the construction of ten nuclear power stations with a total installed capacity of 22 gigawatts, to be in operation by the year 2000. They would provide between 3.5 per cent and 7 per cent of the capacity needed, depending on what assumptions one makes about economic growth. Such a programme would not even fill the gap left by the retirement of the present ageing Magnox and coal-fired stations, let alone make up the deficit of energy due to oil depletion ... It is not even clear that the nuclear programme can be introduced at the rate envisaged. The time taken to build the existing AGRs has been many years longer than planned, suggesting that further development work is needed. The same is true of the American PWRs, unless we are to ignore the Harrisburg accident. In addition, the necessity for im-

provements in uranium mining technology and the demonstration of an acceptable technique of waste disposal, as recommended by the Flowers Commission, may introduce further delays and will certainly increase the costs ... The capital cost of a nuclear power station, if spent instead on energy saving, would save three times more energy than the station would produce in its lifetime.

6. Spent Fuel Transport

Civil nuclear power plants, for reasons discussed elsewhere, are usually sited in remote areas, the only exception to this being the small demonstration reactor at Greenwich. The waste from all the reactors has to be carried to Windscale for reprocessing — thus the spent fuel transport system is the jugular vein of the nuclear power programme, because without it the cooling ponds at the reactors would soon become full, and the reaction would have to be stopped.

Spent fuel transport is not only the most vulnerable phase of the nuclear cycle but also that most susceptible to mass protest. This is the only phase of nuclear power which city dwellers can see for themselves — 8.5 tonnes of highly active spent fuel pass through London each week. In no sense are these 'waste' products — the spent fuel is 99 per cent uranium, the remaining 1 per cent consisting of cesium-137, strontium-90, ruthenium-103 and plutonium-239. The consequences of contamination from a flask accident in social terms have been described by Dr Wakstein as follows:

> An accident to Dungeness fuel at Earl's Court with a west wind could force the evacuation of 80,000 people — including the Royal Family from Buckingham Palace — for 25 years and the loss of at least 20,000 households for times ranging from months to 125 years. Even at the average cost of council housing we are talking about at least £400 million worth of property. We are also talking about creating a ghost town in the middle of the Royal Borough of Kensington and Chelsea.

This scenario assumes a 10 per cent release: should a 100 per cent release occur, London would be sealed off for a century. In nuclear terms, we have become immune to the immensity of the consequences of serious mistakes — the brain becomes accustomed to, and thus does not react to, descriptions of catastrophe. The projection given above was calculated using the UKAEA's own computer programme 'Tirion'. On average, three flasks travel through London each week: the release of the contents of any one of them would paralyse the capital for a century.

It is difficult to know how to grade catastrophes — whether in terms of deaths, administrative chaos, or according to the value of property made unavailable. Certainly, viewed in the light of the last two criteria, a severe accident during the spent fuel transport phase of the civil nuclear power programme would have the most disastrous possible consequences.

This being so, it would seem sensible to examine the methods and safety procedures used in the transport of spent fuel. We shall restrict ourselves to commenting on the movement of fuel across London, as this is the best researched area, and we would recommend 'Carrying the Can' published by the Ecology Party, to anyone interested in the details of waste fuel transportation across the capital.

The fuel which travels across London comes from three Magnox reactors: Sizewell, Bradwell and Dungeness A. The fuel from Sizewell and Bradwell comes via Stratford, and across the North London Line to Willesden Junction. Fuel from Dungeness comes via Croydon through Streatham, Wandsworth Common to Willesden Junction via Kensington, Olympia and Shepherd's Bush. Fuel from all three reactors leaves Willesden on a north-bound train via Wembley Central, Harrow and Wealdstone and Watford Junction, en route for Windscale.

Spent fuel is packed in huge containers, very solidly built. Each flask (loaded) weighs about 50 tonnes and contains up to 3 tonnes of fuel in 200 or more fuel elements. Each container is believed to have cost £250,000. Some flasks, though not all, are tested. The CEGB and the Department of Transport have consistently refused to release the details of the result of tests 'for reasons of commercial security', but we do know that flasks are designed to withstand an impact of 30 mph and a fire of 300°C for 30 minutes. Since we also know that the trains on which these flasks are loaded travel across country at twice the tested speed, we can only speculate about what would happen if a container were derailed into the track of an oncoming train. We know that the temperature of a petroleum fire reaches 1200°C and that of a spent fuel fire 1100°C, and we also know that no regulation forbids the presence of petroleum tankers on the same train as spent fuel. What is not generally realised is that there is a complete absence of impact testing, not even to 30 mph — as all such testing is done on a quarter size scale model. To quote the Royal Commission on Environmental Pollution: 'we were surprised to learn that the tests are conducted only on models, and since the containers travel on ordinary freight trains which may be expected to travel at speeds up to twice that assumed in the tests (with kinetic energy four times that assumed), we were not wholly reassured.'

The CEGB is quick to point out that there has never been a significant accident involving a nuclear flask in transit, although there have been minor derailments in marshalling yards. British Rail statistics show that, between 1968 and 1977, there were 274 collisions involving freight trains. 22 of these were serious and were the subject of inquiries. From what is known about the routes, at least eight of these collisions occurred on lines that carry flask traffic. In the same period there were 2,499 derailments recorded, which is an average of 250 each year. It is a statistical certainty that such an accident will one day involve a spent fuel flask.

In this area, as in other areas of the civil nuclear programme, the various authorities contradict each other. We shall give two examples here, one concerning Safety Related Incidents (SRIs), the other concerning rehearsals with the emergency services.

In January 1978 the Secretary of State for Energy, Tony Benn, asked 'what documents contain a list of all SRIs ... to flasks containing irradiated fuel since 1950?' The answer was:

There is no document listing safety related incidents involving flasks containing irradiated fuel. Such incidents would have been reportable by operators of licensed nuclear sites under the Nuclear Installations (Dangerous Occurrences) Regulations 1965, or in the case of UKAEA, under equivalent arrangements. No such incidents have been reported.

Yet in November 1979 the Parliamentary Under-Secretary at the Department of Transport, Kenneth Clarke stated that since 1976 eight instances of minor contamination above the permitted levels (involving spent fuel flasks) had been reported to the Health and Safety Executive (*Hansard* 8 November 1979 col. 272-3).

The CEGB press briefing of January 1980 stated that:

> The emergency arrangements are regularly rehearsed to improve the smoothness of operation and to gain experience. These rehearsals include participation by police, fire brigades, British Rail staff and expert teams, and they ensure that notification and communication arrangements function correctly. In addition, British Rail have practised the righting of an overturned flask.

The union branch secretary at Battersea Fire Station, which would be called upon to deal with a fire at Clapham Junction, said that his men had never been told that spent fuel was passing through, nor had they done any drill in preparation for a crash involving radioactivity. At a meeting at the CEGB offices in London on 11 January 1980, it was admitted that no full-scale exercise had been carried out on the North London Line in anticipation of an 'incident' and that none was envisaged.

The security of the flasks during every stage of their journey is obviously crucially important — as the Royal Commission commented: 'It is not a question of whether someone will deliberately acquire the waste for blackmail purposes or terrorism, but only when and how often'. Apparently, this view is not shared by the authorities — the Department of the Environment stated in May 1977: 'As the flasks are so stringently designed and their design is assessed by experts under the Secretary of State for Transport's Radiological Adviser, there is no need for any special controls during transport, such as warnings to health authorities and emergency services, and special routes.'

Flasks are easily accessible to the public — witness the famous photograph by Pat Kinnersly showing children playing round a flask at Southminster. There are no fences at Southminster siding, and at Ipswich and Stratford stations anyone can walk up to a flask and touch it.

On 1 November 1979 three men walked up to a flask standing at Platform 9, Stratford Station, and pointed a rocket-launcher at it. The launcher was a model and the objective a photograph, and British Rail commented that it 'is not the job of British Rail staff to apprehend people carrying rocket-launchers'.

Spent fuel must be transported for the duration of the civil nuclear power programme — the only alternatives are to close down the power stations, or to have a reprocessing plant next to each reactor. While we would prefer the former, we recognise that nuclear power may continue to exist for the present. But spent fuel must be routed away from London: 'Carrying the Can' points out that it has already been banned from passing through New York.

> In New York there is a blanket ban on any movement of spent fuel or radioactive waste through the city. This ban did not have to be brought about by public protest or intensive lobbying. It was introduced by the director of New York's Bureau of Radiation Control, Dr Leonard Solon, because he concedes that the consequences of a serious accident are absolutely unacceptable: 'With a massive release we're talking about the potential of thousands of long-term malignancies or cancer deaths and hundreds of prompt deaths.'
>
> Dr Solon has suggested that London should follow New York's example, and he has questioned the safety of the flasks, saying that the possibility of a faulty flask was one of the overriding factors in convincing him of the need for a ban in New York: 'The concern is not with the tests with the flasks or casks that survive … you're concerned about the five out of a hundred or two out of a hundred or one out of a hundred in which there is some kind of engineering defect or imperfection in which some quality assurance procedure has failed. You can't do every cask to destruction'.
>
> One cannot fail to notice the difference in attitude from that of the radiation authorities in this country.

Although the GLC has refused to become involved, on the grounds that spent fuel is a national issue, several local authorities have recently become concerned. Brent Council has called for a public inquiry into the issue and for a moratorium until the outcome of the inquiry. Newhaven has also condemned the transport of spent fuel through London, as has Hackney, who called the consignments 'dangerous and irresponsible' and 'a threat to the lives and health of the people of Hackney and other boroughs'. Camden Council is taking legal advice on the possibility of blockading the transport of spent fuel through the borough, and in early 1981 Wandsworth also called for an inquiry.

We find it invidious that we should have to conclude this chapter with an appeal for spent fuel transport be routed away from London, as this means that other parts of the country would have to tolerate the passage of this poison. The only sane solution is to remove the problem by immediately abandoning the nuclear power programme: until the political will to do this is generated, we can only recommend that the consequences of terrorism or an accident be minimised by removing spent fuel from the capital.

7. Civil Liberties

In a letter to *The Times* on 1 November 1980, Sir Kelvin Spencer, former government chief scientist, points out the difficulties involved in the Cabinet obtaining impartial advice on nuclear matters:

> Sir — Your correspondent Mr Fabian Acker (22 October) asks why those who plead for a 'sensible debate' on nuclear power assume the starting point must be acceptance of the need for it.
>
> This puzzles him. Maybe I could give him the clue. The decades I spent in and around Whitehall, and the freedom I've had since retirement to ponder on the corridors of power, lead me to the following conclusion.
>
> The need for nuclear power has been a conviction of a succession of Cabinets since the UKAEA was set up in the 1950s when I was Chief Scientist at the Ministry of Fuel and Power. Cabinet members haven't the time to study complex issues. They must rely on advice from what they hope are competent and unbiased sources. In the nuclear field there are four main sources: civil servants, the staff of UKAEA, industry, and the financial world.
>
> Civil servants with any deep knowledge of nuclear complexities left government service for UKAEA or industry long ago. The pay is better, and a job in UKAEA or on a nuclear contract is screened from the financial axe. Hence the advice civil servants give to ministers is what they themselves get from the other three sources.
>
> It needs no crystal ball to see what advice UKAEA will give civil servants. Advice from industry will be much the same. A firm

that gets contracts for anything to do with nuclear power gets paid for good work, and paid too for putting right work not so good. Their interest is to keep the nuclear bandwagon rolling.

International finance is deeply committed to the mining and marketing of uranium ore. Massive investments have been made in the belief that there will be an expanding market for it. That depends on there being a continuing programme of nuclear power stations. Advice from the City can therefore hardly be relied on for impartiality.

But government is faced with growing opposition from the general public. How better could this be countered than by widespread debate made sterile by the myth of *need*.

The alleged need is given plausibility by playing down the alternatives — the so-called renewables such as wind, sun, waves, tide, geothermal heat and the like. Public funds for developing these are administered by the Energy Technology Support Unit. This is located at Harwell, the centre of nuclear technology. This tilts the scales against them from the outset.

As for using energy more efficiently by such measures as thermal insulation, the present government has reduced financial incentives for this.

Starving alternatives to nuclear energy while generously financing the latter has the inevitable consequence of helping to promote the fiction that *only* nuclear energy will 'Fill the gap'. Public debate against this background helps to spread the myth.

Equally worrying, in the context of the decreasing range of official information available, is the restriction of the public inquiry system. When British Nuclear Fuels Limited applied to build a large reprocessing plant at Windscale in 1977, they omitted to mention that they would need to increase the amount of water they take from the beautiful lake of Wastwater from four million to seven million gallons a day. Residents of Wasdale Head, in particular the farmers, were distressed by the proposed plans for a weir which would raise the level of the lake by several feet, causing flooding and the destruction of agricultural land. Since the plant has not yet been constructed, the point should surely have been raised at the original inquiry; as it is, it has necessitated a whole new public inquiry. This is very expensive and unpleasant for the local community who are trying to protect their livelihoods and their lives — less so, of course, for BNFL, who apparently have unlimited amounts of taxpayers' money at their disposal to present a case which should have been dealt with earlier. The Windscale Inquiry confined itself to discussing the immediate planning application — discussion on the wider issues of nuclear power was not permitted. This precedent, which should be a matter of concern to anyone committed to safeguarding civil liberties, and to organisations such as the Town and Country Planning Association, will be vigorously opposed at the inquiry into the PWR application at Sizewell. Opposition groups have demanded that the CEGB publish its studies into the PWR's safety at least six months before the inquiry, so that they can be challenged by outside experts — and such groups are also pressing for public funding at the inquiry, to the same extent as the nuclear industry. But expectations of impartiality at Sizewell are already being undermined by the exploratory work that is currently taking place on the projected site.

The investment in nuclear power is so huge, and the power of the corporations involved so massive, that the rights of citizens all over the world are being threatened. A court in Japan dismissed arguments that the constitution gives people the right to live in 'a good and natural environment' and ruled that a nuclear power plant could become operational. But in Austria, a public referendum decided against permitting the use of a nuclear power plant which had already been built; while in the US, the aftermath of the Three Mile Island accident has ensured that all new nuclear projects have been frozen, and the fast breeder programme abandoned indefinitely.

The nuclear power process is critical at every stage — there is neither the time nor space for industrial democracy. A nuclear power programme inevitably tends to lead to an increase in totalitarianism and to an erosion of civil liberties; the larger the programme, the greater the erosion. Our civil liberties are threatened because of the military connections of nuclear power and the potential for terrorism, especially in terms of plutonium; because of the extensive dispersal and long-term effects of the lethal and genetically mutating waste produced by the industry; and because of the critical nature of the nuclear process which, once started, cannot be left untended or unguarded at any point. When, in 1977, workers at Windscale staged an unofficial strike for danger money (a thousand workers had been laid off because they could not be issued with protective clothing), the Secretary of State for Energy announced that the army would be brought in with supplies of nitrogen to keep the plutonium store atmosphere inert. The criticism here is not of Mr Tony Benn, whose immediate duty was clearly to ensure the safety of the plant, but of the process itself, which as well as imposing difficult and dangerous working conditions, totally overrides the rights of the workers. At Cap de la Hague in France, where workers and residents have been active in demanding safer working and environmental conditions, an 80-room police barracks was built at the same time as a new reprocessing plant.

It is not only the workers' rights that are at risk; in 1976 the Labour government passed the Special Constables Act, which licensed a special police force for the nuclear industry. These policemen are armed and authorised to shoot *on suspicion* as well as having rights of entry and search also on simple grounds of suspicion. They are under the control of MI5 which, as the Blunt affair showed, is answerable only (if at all) to the Home Secretary; the public has no right of appeal to the courts or to the ombudsman.

The creation of this nuclear police force may be rationalised as being necessary to prevent the theft of plutonium or of similar radioactive substances for bomb making, blackmail or other terrorist purposes — but security, even with the special police, does not seem to be very efficient. In November 1979 three members of the Freedom of Information Campaign staged a rocket-launcher attack on a container of radioactive waste at Stratford station in the East End of London. The trio bought platform tickets and waited for the train, not bothering to hide their rocket-launcher, which was in fact a model. Had it not been, and had the demonstrators been terrorists, the attack would have dangerously contaminated 1.75 square miles of London for the next 125 years. British Rail spokesmen maintain that nuclear material is moved under the strictest security, but recognise that it is impossible to guard the whole network against saboteurs. One said, 'The traffic has been passing for many years now without incident.' Nevertheless, it cannot be denied that the Freedom of Information Campaign demonstration graphically highlighted the inadequacy of current safety arrangements during the transport of nuclear waste.

An armed, self-regulating police force has been imposed on the public; it is demonstrably not doing the job for which it was set up; and that job may well be impossible without the introduction of a completely totalitarian state. These observations underline the confused, inadequate and ostrich-like thinking which, along with the power and secrecy with which it operates, seems to be a hallmark of the nuclear industry in Great Britain. These characteristics had already been noticed by members of the Royal Commission on Environmental Pollution, headed by Sir Brian Flowers, in 1976, the same year in which the special police force was created. A member of the commission, Baroness White, said in the House of Lords: 'Some of us ... were often depressed by what we felt was the blinkered outlook which could preclude adequate consideration of matters not directly in the line of vision. We were depressed by a certain rigidity of mind, and in some cases by what I have called the inpenetrable complacency of those in high places. A Micawberish attitude does not go well with a substance as potentially risky as plutonium' (*Hansard*, 22 December 1976). Three years later, in April 1979, the then Secretary of State for Energy, Tony Benn, gave a clearer warning. In a speech entitled 'The Democratic Control of Civil Nuclear Power', given at the inaugural meeting of the Science, Technology and Society Association, he listed four principal areas of danger for democracy:

1. The close connection with military policy in regard to the development of nuclear weapons, which were developed secretly in the UK without parliamentary knowledge or approval.
2. The link between the military and civil use of plutonium, making it available for proliferation and terrorism.
3. The wide gap between expert understanding and public knowledge of nuclear matters, which is used to justify the exclusion of laymen, even MPs and ministers.
4. The high rate of expenditure which creates powerful vested interests in those industries which live on those budgets.

He pointed out that the implication for democracy, in the use of a technology as complex as that of nuclear power, is 'that the information for democratic control is not as readily available as it should be'. **Having formerly been Minister of Technology, then Secretary of State for Energy, he supported this statement with examples:**

1957. The Cabinet was not informed of the catastrophic nuclear waste accident which permanently wiped out a huge area in the USSR. However, the UKAEA knew of this disaster.

1968. The Cabinet and the Minister of Technology (Tony Benn) were not informed that 200 tons of uranium had been hijacked from Euratom. However, the UKAEA knew.

1969. When the corrosion of the Magnox fuel cladding was discovered at Windscale, with grave effects of overheating and radioactive release, Mr Benn as Minister of Technology was informed, but pressured to keep quiet and not cause alarm.

1970. Rio Tinto Zinc signed a uranium contract with Namibia (South Africa) on behalf of the UKAEA before informing the Cabinet.

1972. The Vintner Report, a study on thermal nuclear reactors carried out by the Department of Industry, was withheld from the Commons Select Committee on Science and Technology 'for considerations of commercial confidentiality'.

1976. The major radioactive leak at Windscale was not fully revealed to Mr Benn as Secretary of State for Energy. He subsequently instigated the quarterly reports on nuclear failures and 'incidents' which must be presented via the Health and Safety Executive to the Secretary of State. Mr Benn added that he felt that the public had not been clearly informed that the process of vitrification of waste had not been fully perfected.

Finally, Mr Benn listed the particular points of vulnerability in a high-technology society:

1. Strikes.
2. Technical disasters.
3. Natural disasters — earthquakes etc.
4. War.
5. Fear (of these above) breeds insecurity which in turn breeds loss of freedom (e.g. the constant policing and barbed-wire security which surrounds the experimental FBR plant at Dounreay).
6. Opponents of nuclear power, or students questioning the

energy system, could be observed and put on record for 'thought-crimes' (as in *1984* or Watergate).

Mr Benn ended his speech with a question that might sound melodramatic if events had not begun to give it serious import. Emphasising the lesson we should have learned from the Industrial Revolution, that technology cannot be discussed in isolation from the values of society, he posed the question: 'Could we back into a police state because of high technology?'

The story of Karen Silkwood, who was an employee of Kerr-McGee in Oklahoma, lends weight to Mr Benn's question. It was first told in this country by Jim Garrison, in the *Ecologist*, November 1979. Kerr-McGee had recently turned from manufacturing parts for the oil industry to producing the plutonium fuel rods for an experimental FBR being built by the Westinghouse Corporation, on a multi-million dollar contract. Karen Silkwood was trained in physics and believed in the benefits of nuclear power; she was hired as a laboratory technician to test the quality of the plutonium fuel rods produced at the plant. Once there, however, she began to complain about the health and safety conditions at the plant. Untrained workers were put on plutonium fuel rod production, others were sweeping up plutonium in the dust from the workshop floor, and the management ordered workers to continue working in highly contaminated areas while clean-up crews attempted to decontaminate around them. The United States Atomic Energy Commission health and safety regulations were being violated on all sides. Even worse, from the public's point of view, quality control standards of the fuel rods themselves were being grossly violated. She told the Oil, Chemical and Atomic Workers International Union that she could get 'documented proof that Kerr-McGee officials were knowingly "doctoring" with magic marker the safety inspection X-rays' which, by USAEC regulation, had to be taken of every fuel rod, to ensure that it was not leaking radiation through faulty welding. The USAEC demanded perfectly welded fuel rods because, if defective, they could cause a disturbance in the flow of the liquid sodium used to cool the reactor core of the fast breeder. If the sodium is blocked and prevented from reaching the portion of the fuel it is meant to cool, the fuel rods can overheat, leading to accidental releases of radioactivity. Under worst-case conditions, faulty fuel rods can lead to a meltdown of the reactor itself.

Tony Mozzochi (vice president of the union) was later to testify in a federal court that the charges made by Karen Silkwood 'were the most serious charges I have heard in my trade union experience'. Karen Silkwood was instructed by Mozzochi to return to her job in Oklahoma and secure copies of the doctored X-rays, along with as much evidence as possible to substantiate the 40 other charges she had made concerning Kerr-McGee health and safety violations.

What followed has the quality of a nightmare. Under the cover of an interested local reporter, the Oklahoma State Police gathered all the information they could from Karen Silkwood herself (who thought they were genuinely interested in the issues as reporters) at informal encounters in restaurants. The state police then put FBI agents on surveillance of the case. Shortly afterwards, Karen Silkwood found she was radioactively contaminated. There was no indication that this had originated from her working, but the plant decontaminated her. Over the next few days this pattern was repeated; the only contaminated areas in her working environment were those she had touched herself. Finally, Kerr-McGee themselves ordered that she should fly to Los Alamos to a special government laboratory which would determine the extent of her internal contamination. They added that, unfortunately, she would thereby miss the important meeting to negotiate workers' safety contracts she had been working for. Meanwhile, in her absence, her life was thoroughly inspected; letters, documents and memoranda were turned over, not to the medical decontamination team, but to the chief of Kerr-McGee security.

The Kerr-McGee team was then joined by inspectors from the Atomic Energy Commission. They discovered that the food in her refrigerator had been radically contaminated with plutonium. The contents of her whole apartment were sealed in the drums used to dispose of plutonium waste. But although Karen Silkwood was scheduled to stay in Los Alamos for more tests, she flew back to Oklahoma and told her union supporters that she still 'had the documents' and was prepared to pass them on. She made arrangements (by telephone) to meet an investigative reporter from the *New York Times*, together with union representatives, on 13 November 1974. She never arrived. Her contacts telephoned and found that the phone she had called from was 'out of order'. They then drove off to find her, and discovered instead an accident to her car in which she had been killed. Her body was gone, and her car had been taken away — but not by the highway patrol, as is normal. The local car-wrecker service who had been called — by whom? — who knew about the accident before the highway patrol? — said that he had been ordered by Kerr-McGee officials, dressed in radiation suits with face-masks, to inspect it 'for radiation'. According to the wrecker, the same officials took documents from the car and began reading them aloud to each other. Nobody remembers whether these men took away the documents. Later, the local police came to the garage and also took some things from it; afterwards the garage-owner and wrecker, Ted Seabring, removed all the remaining items from the car and sealed them in a cardboard box. When this was opened later, in the presence of witnesses, it was found to be empty. Within a few months of her death, the case of Karen Silkwood was 'officially' closed — except to the many people who remained unsatisfied. Answering official enquiries by her parents, the National Organisation for Women, and her union, the FBI inspector said that if they wanted every case to be solved, 'they watched too much television'.

However, in November 1976, it was legally charged that Kerr-McGee was liable for the plutonium contamination of Karen Silkwood. Other charges included a conspiracy to cover up a deprivation of the equal protection of the law, and to cover up a deprivation of the right to equal enjoyment of the privileges and immunities of all those persons (of whom Karen Silkwood was one) who reported violations of the federal Atomic Energy Act by the Kerr-McGee Corporation. At the final hearing, after considering depositions by independent investigators that Karen Silkwood's car must have been forced off the road, it was concluded in evidence at the Oklahoma Federal Court by Dr John Gofman that Karen Silkwoood was 'married to plutonium' because federal standards are 480 times too lenient. Dr Edward Martell, an environmental radio-chemist, testified that existing radiation exposure limits are 'misleading and inadequate and have not been reduced because of the government's vested interest' in nuclear power. He called standards in the US and abroad 'meaningless' because, contrary to official policy, there is no safe limit for exposure to low-level ionising radiation. Dr Karl Morgan, often referred to as the 'father of health physics' because of his role in the setting of acceptable standards for radiation releases in nuclear facilities, testified that Kerr-McGee training manuals showed a 'callous attitude' towards the safety of its workers. He pointed out that the training manuals made no mention of the fact that one could contract cancer from radiation exposure, and that because of their refusal to recognise the dangers of radiation, the company and the workers were in violation of safety regulations.

At the end of the case, the jury at the Oklahoma Federal Court awarded ten and a half million dollars in actual damages and $500,000 in personal injury damages to Karen Silkwood's three children.

> In charging the jury, Federal Judge Frank G. Theis directed them to define 'physical injury with regard to plutonium as ... nonvisible or indetectable injury ... to bone, tissue or cells'. The implications of this are profound, for it establishes *legally* that plutonium is a 'dangerous material' and causes 'physical injury'. This means, on the other hand, that nuclear materials are so dangerous that nuclear facilities are under special restraint to prevent the escape of any of the material, whether intentional or otherwise; and on the other hand it means that workers and members of the public are now entitled to claim damages due to the operation of nuclear facilities if they can demonstrate that their sickness is attributable to radioactive releases coming from the plant involved. (*Ecologist*, November-December 1979).

When charging the jury, Judge Theis stated that they did not have to find that Kerr-McGee deliberately contaminated Karen Silkwood; the fact that plutonium had been allowed to escape from the plant was reason enough to award damages.

Two very important points stem from this case. Firstly, that workers and the public in the US can claim damages against the release of radioactive substances from nuclear plants; and secondly, that even though Kerr-McGee was not found guilty of deliberately contaminating Karen Silkwood, nevertheless the company still appealed against the judgement. This is very unusual on the part of a large company known to have been guilty of safety violations, and suggests that, if Kerr-McGee was convinced of its innocence, someone else was guilty. Certain odd features of the case spring to mind. Was the company quite innocently testing and decontaminating her in the days before she had prepared her dossier for the union and the *New York Times*? If not, was the company subjecting her to this unpleasant process as a further harassment? If harassment, why contest the damages? And, in the end, who *did* contaminate the food in Karen's refrigerator? Could someone even more powerful than Kerr-McGee have been involved, perhaps someone who had originally awarded the huge plutonium contract to this inexperienced firm? Someone introduced the state police and the FBI to the case before it was a 'case'; someone removed her body and the documents she was carrying, before reporters or local police could reach the scene of her death. If it was Kerr-McGee who did these things, that is bad enough; if, as their appeal suggests, it was someone else, that is even worse.

Who then are the terrorists? The workers, who are trying to protect their health and safety at work? The employers, who should be protecting their workers and the public while trying to make a profit? Or some even larger agency, not necessarily from the nuclear industry, who is manipulating this potentially deadly process for its own ends? The case of Karen Silkwood is an indictment, not simply of the United States Atomic Energy Commission, nor only of Kerr-McGee, but of the nuclear process itself as a part of society. Its hazards are too obvious, too deadly, too open to abuse, for a sane society to accept.

The directive given by Judge Theis in the Karen Silkwood case is clearly a major step in re-establishing civil rights in the nuclear context; but we must remember that it only applies under US law. In 1978 a group called Justice, the British section of the International Commission of Jurists, published a report called *Plutonium and Liberty: some possible consequences of nuclear reprocessing for an open society*. Discussing the legal implications of discharges of radioactive substances, they foresee great difficulty in resolving the question of damage, and damages, for the whole nation:

> As we understand it, neither power stations nor nuclear reprocessing plants are designed to release no radioactivity whatever: they are — and under existing and foreseeable constraints, both of

technology and cost, they have to be — designed to release 'low-level', 'insignificant' or 'negligible' amounts of radioactivity into their environment.

In this respect, such installations do not differ from others in which dangerous materials are handled: many other industrial processes discharge small quantities of noxious substances into the environment. Nor are radioactive substances unique in taking a long time before they cease to be harmful to life, or in being subject to reconcentration in biological food chains and other environmental pathways: organochlorides such as DDT produce both these effects.

However, there would appear to be one qualitative difference between radioactive and most other noxious substances (some chemical mutagens being an exception). As we understand it, the weight of radio-biological evidence supports the view that there is no 'threshold' below which ionising radiations are totally harmless: however small the dose to which a large population is exposed *some* individuals within that population will suffer somatic ill-effects (such as cancer or leukaemia) or genetic damage, though it may take years (and, in the latter case, generations) before the ill-effects become manifest.

It is, of course, common practice in all modern industrialised societies to accept, tacitly or explicitly, a small cost in health, or even life, as the price of the benefit accruing to society generally from industrial processes. Road transport, which in the United Kingdom alone accounts annually for over 7,000 fatalities and over 350,000 injuries (more than a third of them serious), is one of the more obvious examples. But an essential corollary to this social acceptance of a hazard is that society must ensure that those for whom the hazard is realised are compensated as fully as they can be, either by those who create the hazard, or by those who enjoy the concomitant benefit. In the case of radioactive substances, there appear to be some substantial obstacles in the way of achieving that objective.

First, there is the 'background' radiation to which everyone is unavoidably exposed, and which takes its 'base level' toll in somatic disease, and genetic damage, among individuals whose statistical distribution within the population may be broadly predictable, but who cannot be identified. Secondly, radioactivity is not the only cause which can produce or precipitate these forms of damage: other agents, of which some may not have been identified, may have the same effect — as in the case of lung cancer, which may follow the inhalation of plutonium oxide, or of cigarette smoke, or of exhaust smoke from diesel engines.

Accordingly, we foresee difficulties, which may in practice

prove insurmountable, in the path of any victim wishing to prove, on the balance of probabilities, that his affliction was caused by 'low-level' radioactivity discharged from a nuclear installation, rather than by some other agent ... We are therefore driven to the conclusion that the operation of any large-scale nuclear power programme, with the inevitable increase in the total dose of ionising radiation to which the population as a whole will be exposed, will lead to a proportionate increase in uncompensated, and uncompensatable, disease, death, and congenital malformation among individuals who form part of that population. It is not for us to judge whether that is a consequence which society should or should not accept. All we can say is that it would be an injustice: whether it is an 'acceptable' injustice is for others to determine. Those others will include some, but by no means all, of the potential victims.

The risk of terrorism throughout the country; a police state; a diseased population unable to claim just compensation; and the genetic weakening of future generations. Are we really unbalanced enough to accept these as part payment for what has been called the Faustian bargain? If we had been given the choice, that is. Had we been given the open choice, the financial burden and the accident risks alone would have been enough to deter all sane citizens; to add these threats to our civil liberties must surely settle the question on equally grave moral grounds. The Royal Commission on Environmental Pollution came to the same conclusion:

> Our basic concern is that a major commitment to fission power and the plutonium economy should be postponed as long as possible, in the hope that it might be avoided altogether, by gaining the maximum time for the development of alternative approaches which will not involve its grave potential implications for mankind.

8. Doing Without Nuclear Power

It now only remains to put nuclear power in the context of the energy needs of this country. It should be remembered that electricity — generated by coal, oil and nuclear methods — accounts for between 8 and 12.5 per cent of our total energy consumption. Electricity generated by nuclear power at present supplies 2 per cent of our total energy consumption. Nuclear reactors and their associated plant, with their constant leakage of radioactivity into the environment, with their insoluble waste control problems, with their devastating potential for destruction, both in terms of accidents and of the theft and misuse of plutonium, with their requirements for totalitarian security procedures, and with their astronomical costs — these machines contribute 2 per cent to our energy needs. The present government has plans to build ten more nuclear stations during the coming decade, at an estimated price of £1,500 millions each. At the end of that programme, nuclear power would only supply 3.5 - 7 per cent of our energy needs, depending on the eventual growth in electricity demand. Never has so little been achieved at such a price.

Forecasts of future energy growth — upon which the CEGB bases its power station construction programme — are changing rapidly. The assumptions on which the plans for ten new reactors were based are now already hopelessly out of date. As the report *A Low Energy Strategy for the United Kingdom* by Gerald Leach et. al. points out: 'After many years of rapid growth, at about 7 per cent, slowing to 4 - 5 per cent per year in 1973, electricity consumption has been roughly static, so that in 1977 it was still at the 1973 level. Power station programmes based on early, optimistic forecasts of high growth have therefore produced an embarrassingly large plant over-capacity; a position that will worsen as the power stations now under construction come on stream.' Recent figures issued by the Department of Energy show an actual decline in consumption — the July 1980 copy of the Department's publication *Energy Trends* shows that, on a seasonally-adjusted and temperature-corrected basis, Britain's energy consumption in March-May 1980 was 7.7 per cent below the consumption in March-May 1979: this trend continued throughout 1980. Although this figure refers to total energy consumption and not only to electricity, it must be becoming obvious to CEGB statisticians that the period of no-growth in electricity demand from 1973 until the present can no longer be regarded as a temporary pause in the otherwise smooth slope of continuously increasing growth, but has become a feature significant to future planning. Indeed, the position of the CEGB in the nuclear power debate is becoming increasingly ambiguous

— on the one hand it is heavily committed to nuclear power both in principle and practice; on the other hand it is realising that if the political decision massively and rapidly to increase the nuclear power programme is executed, it will have to pay for the excess generating capacity so produced, without hope of financial help from the government that created the problem.

The stabilisation of electricity consumption should come as no surprise to the politicians, nor should it cause them undue worry. The correlation between increasing energy consumption and an increasing gross domestic product was for many years seen by economists as an immutable fact. But it is becoming clear that it is possible to have healthy economic growth without a corresponding growth in energy consumption — as the emphasis in society shifts from manufacturing industry to service industry with its lower energy intake (the silicon chip is not a great consumer of electricity), this becomes inevitable. On another level, for the last 25 years energy consumption in the housing sector and personal income levels have risen in parallel. But this obviously cannot continue indefinitely — at what income level do houses become so hot that they are uninhabitable? Increasing wealth and increasing energy consumption do not permanently go hand in hand.

The previously mentioned report, *A Low Energy Strategy for the United Kingdom*, is a meticulous and scholarly study which demonstrates how, even assuming the economic growth which seems so elusive, we can use the same amount of energy in the year 2020 as we do now. The cornerstone of the method by which this may be achieved is energy efficiency — for even at a time of increasing scarcity, difficulty of production and expense in all energy areas, we are still being profligate in our consumption. After three decades of easy availability we have got into the habit of simply increasing our demand, instead of examining ways of maximising efficiency. We have become used to thinking that to build another power station is easier than insulating the loft — we have been using bigger and bigger standpipes to fill the energy bath, instead of using the bath plug. We have somehow assumed that the constant increase in power supply has been without serious economic or ecological consequences — this assumption, in terms of an increase in nuclear power generation, can no longer be made.

The political decision — energy efficiency or more nuclear power — is a difficult one. In effect, it is a choice between creating a sustainable, safe and low-technology energy future, and jeopardising the future (in their present form) of several large industrial and technological organisations which are heavily committed either to nuclear power itself or to a constantly increasing energy demand. The correct decision will also imply the abandoning of the exciting and fascinating field of nuclear physics in which Britain has traditionally played a significant role. But there are great political gains to be made from deciding in favour of energy efficiency. At a time when the unemployment statistics are perhaps the single most difficult set of figures that the government has to deal with, it is worth remembering that the lower the energy technology, the more jobs are created by its implementation. It has been estimated that half of the cost of one new nuclear power station (assuming the cost to be £2,000 million) would insulate the loft, walls and windows of every house, office and factory in the South of England. A massive insulation programme — the logistics would be similar to the recent nationwide conversion to North Sea Gas — would provide tens of thousands of jobs in factories across the land. If the power stations are built instead, each one will provide 150-200 highly-specialised jobs, and if, as is being proposed, American-designed reactors are used, large numbers of extra jobs in the R & D laboratories will not materialise either.

Energy efficiency, in terms of insulation, makes obvious and immediate sense. As Sir Martin Ryle pointed out in a quotation used earlier, if the money that would be spent on building a new nuclear station were spent instead on insulation, three times as much energy would be saved as the power station would generate in its lifetime. An insulation programme would buy us time — time to debate the issues of power generation in parliament (which, incredibly, has not yet discussed the nuclear expansion plans) and in public, and to evaluate all the available alternatives. Yet almost the first thing that the present government did on taking office was to decimate the insulation grants available to industry and to householders — an incomprehensible decision, presumably designed to protect and increase the profits of the nationalised energy corporations. A few months later, we were told that electricity prices were increasing because of insufficient consumption, a clear admission of excess generating capacity. Nevertheless, the ten new nuclear stations planned for the next decade have not yet been cancelled.

It is not within the scope of this book to discuss alternative methods of electricity generation in detail. It has often been stated, as an argument in favour of nuclear power, that nuclear generation would reduce the power of the coal miners in industrial disputes. This is not a democratically acceptable argument. Many of the alternative technologies reflect the growing view that we should not put all our power eggs in one basket of whatever kind, but rather that we should decentralise and despecialise the whole system, using whichever technology is suitable for the needs and resources of the area served. Solar, wind, wave and geo-thermal sources may all have their part to play in the energy supply systems of the future, and it is certainly true that, had they received a fraction of the nuclear research budget, they would be technically vastly more advanced than they are at present. One technology which has been working successfully for years, both on the Continent and at Battersea, is that of combined heat and power stations. A normal power station is about 30 per cent efficient — the remaining 70 per cent of the generated heat goes up the chimney or into the cooling water; this is true of all methods of power generation. CHP stations make use of this waste heat to heat houses and factories close to the station, thus increasing the overall efficiency of the power station from about 30 per cent to about 70 per cent. CHP provides 20.7 per cent of the power used in West Germany, 29 per cent in Belgium, and 35.8 per cent in Denmark.

These and other alternative technologies have been fully documented elsewhere. Recently in the *Guardian* (4 December 1980) Anthony Tucker pointed out that but for the quirks of fate, the photovoltaic cell might have been developed as an energy source far superior to — and cheaper than — nuclear power. The relevant quirks of fate were the demands of the US space programme. Tucker concludes:

In analysing the situation, two factors dominate: the costs of nuclear power have risen substantially over the past decade and are expected to continue to rise; the costs of photovoltaic cells have fallen from 1,000 dollars per peak watt in 1970, to 10 dollars in 1980. On the basis of present technical developments and the economies of bulk production, a further, if smaller, fall (down to one cent a peak watt) is expected by 1990. A forward look examined critically by Bockris and others sees solar voltaic energy becoming increasingly cheaper than nuclear well before the turn of the century. It is the period beyond that which should be concerning us now.

We ought to be able to perceive the effects of costly commitment and illogical bias in energy conversion investments, and the narrowing tunnel into which we are driving ourselves. We do not have to be a visionary to recognise that any light at the end of the nuclear tunnel may be that of final oblivion. The cruel fact that this aspect of the problem is not balanced by any promise of a solution to other aspects of the energy problem makes the situation extremely gloomy. The important thing is that, as in spaceflight, the solutions are at hand if we will invest in them, and they will be at hand in time if we invest rapidly enough. The tragedy is that the governments of the industrialised world, distorted by commitment and dominated within their administrative structures by matters of defence and the bias created 40 years ago, can see only backwards even when they try to look forward. In any case in politics a month is a long time; in technological transition 50 years is a minimum period of change. The paradox and the mismatch can take us to disaster.

Otto Frisch, who was at one time director of High-Energy Physics at the Cavendish Laboratory in Cambridge, agrees with Tucker in reminding us of the origin of all energy:

Uranium-burning stations, plutonium breeders, even reactive lithium in hydrogen power stations will always be dangerous to human beings. I look forward to a world where men no longer depend on fossil fuels — not on coal, wood, oil — nor uranium and hydrogen. It may take a long time, but man must break the habit of using whatever happens to be lying around to meet his growing energy needs. In the end he will be compelled to stop his wandering into such thickets of danger and to turn back to the original source of all energy. He must turn to the sun. I am sure, finally, there will be dramatic advances in development of techniques for storing and using this natural font of power.